DEVELOPMENTS IN LIGHTING—1

DEVELOPMENTS IN LIGHTING—1

Edited by

J. A. LYNES

B.Sc. (Eng.), M.A., C.Eng., M.I.E.E., F.C.I.B.S.

Principal Lecturer, School of Architecture,
Plymouth Polytechnic, Plymouth, Devon, UK

APPLIED SCIENCE PUBLISHERS LTD
LONDON

APPLIED SCIENCE PUBLISHERS LTD
RIPPLE ROAD, BARKING, ESSEX, ENGLAND

British Library Cataloguing in Publication Data

Developments in lighting.
1.
1. Lighting
I. Lynes, John Alexander
621.32′2 TH7703

ISBN 0-85334-774-3

WITH 71 ILLUSTRATIONS

Printed in Great Britain by Galliard (Printers) Ltd, Great Yarmouth

PREFACE

As the insights of the sixties congeal into the orthodoxy of the seventies the forefront of lighting research and practice becomes harder to recognise. The essays in this book are eight individual reports from the frontiers. Together they identify the emerging growth points in the study of lighting today.

In the first chapter Kit Cuttle records the birth-pangs of 'the new photometry', and concludes that some of the classical problems of lighting theory are still unresolved. One problem in particular, the prescription of lighting levels, has been re-emphasised by the energy crunch. This is pursued in subsequent chapters by Gary Yonemura and by Peter Jay.

Yonemura presents a persuasive case for using the modulation transfer function (MTF) in the study of visual performance. His own experimental results, which throw into question the case for higher and higher lighting levels, are of profound importance in their own right. They also offer a foretaste of the effectiveness of the MTF approach. Jay too strengthens the links between lighting theory and recent advances in other branches of vision research. The implications for lighting practice are beginning to emerge.

Peter Boyce, in Chapter 3, shows increasingly subtle techniques for eliciting subjective judgements of electric lighting. This new approach may well become the 1970s' main contribution to lighting research, but it is no substitute for the insight of the gifted designer. Such insight is offered in Chapters 5 and 6 by Jack Waldram and David Clark, respectively.

Designed Appearance Lighting originated over 20 years ago. As Chapter 5 shows, it continues to evolve and to stand the test of time whereas more recent essays in luminance design have sunk without trace.

Chapter 6 by David Clark is a plea for visual literacy, and shows the contribution that perceptive engineering can make to the outdoor environment by day and by night.

The energy crunch has forced lighting designers to take a fresh look at the engineering of daylight; in Chapter 7 Jimmy Longmore discusses new techniques for window design which are developing in response to the fresh demands.

Finally, in Chapter 8, George Woodford distills the lessons of a career devoted to devising and interpreting town planning controls based upon the obstruction presented to daylight by adjacent buildings. The discussion concentrates on daylighting, but its implications must be faced by code-makers for electric lighting and for every other facet of the built environment.

The eight independent contributions were never intended to add up to a coherent philosophy, but one theme that recurs again and again is the need for a sense of direction. Without this, research is fruitless, legislation is self-defeating, and lighting becomes arbitrary and capricious. No message could be more timely today.

For me it has been a happy experience to edit this book. I am grateful to the authors and publisher for making it possible. For its virtues they must take the credit. For its defects I alone am responsible.

CONTENTS

LIST OF CONTRIBUTORS

P. R. Boyce

Research Officer, The Electricity Council Research Centre, Capenhurst, Chester CH1 6ES, UK.

D. McH. Clark

Chief Electrical Engineer for the City of Norwich, City Hall, Norwich, UK.

C. Cuttle

Senior Lecturer in Environmental Studies, School of Architecture, Victoria University of Wellington, Private Bag, Wellington, New Zealand.

P. A. Jay

Senior Partner, Peter Jay and Partners, Consulting Engineers, 1 Penfold Place, London NW1 6RJ, UK.

J. Longmore

Senior Lecturer in Environmental Design and Engineering, Bartlett School of Architecture and Planning, University College London, Wates House, 22 Gordon Street, London WC1H 0QB, UK.

J. M. Waldram

Formerly Senior Lighting Engineer, Hirst Research Centre, General Electric Co. Ltd, Wembley, UK; Consulting Lighting Engineer, 247A Hills Road, Cambridge CB2 2RP, UK.

G. P. Woodford

Formerly Senior Planning Officer, Department of the Environment, London, UK; 26 *rue Lamartine, Pont l'Abbé* 29-120, *France.*

G. T. Yonemura

Research Psychologist, Sensory Environment Program, Environmental Design and Research Division, Centre for Building Technology, Institute of Applied Technology, National Bureau of Standards, Washington D.C. 20234, *USA.*

Chapter 1

THE NEW PHOTOMETRY

C. Cuttle
M.A., F.C.I.B.S., M.I.E.S. (New York)
Senior Lecturer in Environmental Studies, School of Architecture, Victoria University of Wellington, New Zealand

SUMMARY

Several new measures for prescribing the quality of indoor lighting have found acceptance quite recently in various parts of the world. These include Equivalent Sphere Illuminance, the Vector and Scalar concepts of illumination, Mean Cylindrical Illuminance and room surface illuminance ratios.

The author seeks to set their measure into context by examining the thinking and some of the circumstances that led to their introduction. It is suggested that while the concepts that govern current lighting practice are substantially different from those of the past, these concepts derive from a few basic objectives that can be represented as three mainstreams of thought that run right through the development of lighting practice.

The basic concepts of photometry have not changed in almost a century; the Primary Standard, Metrication and the International Vocabulary[1] have brought new units and terms, but no new concepts.

The New Photometry comes from a new sense of geometry. The change has not come about in an orderly manner; if a well-informed lighting practitioner had been asked at any time until perhaps ten years ago to name the main developments he foresaw, almost certainly he would have predicted that luminance would replace illuminance as the most used lighting quantity, and that lighting design objectives would be stated in terms of brightness. Clearly, this has not happened. What has happened is that there have been some extraordinary developments of the concept of illuminance, so that now it is used in a variety of forms and for a variety of purposes not envisaged by its originators.

This chapter aims to explain these developments by describing the circumstances that led to their introduction.

In the first place, it is the author's view that the main developments in the theory and the practice of lighting that have occurred in the past can be attributed to at least one of three mainstreams of thought. Each of these can be identified as having a principal objective, and they will be referred to as:

1. Indoor daylight.
2. Performance-based standards.
3. The design of the visual field.

They are reviewed briefly in the following pages, in which it is shown that each had powerful protagonists and achieved notable measures of success, but yet failed to achieve the widespread acceptance that was sought.

THE IDEAL OF INDOOR DAYLIGHT

At first, electric lighting was seen to be so superior to its forerunners that every aspect of its appearance was greeted with rapture. Consider this comment from a letter from Sir W. G. Armstrong, who in 1881 wrote to *The Engineer* describing the installation of Swan lamps in his house:[2]

'In the passages and stairs the lamps are for the most part used without glass shades, and present a very beautiful and star-like appearance....'

It required a few years of familiarity with the new sources for the development of some basis for evaluating their applications. The following quotation is taken from the second edition of *The Art of Illumination* by Dr Louis Bell, which appeared in 1912.[3]

'In domestic, as in other varieties of illumination, two courses are open to the designer. In the first place he can have the whole space to be lighted brought uniformly, or with an approximation to uniformity, above a certain brilliancy, more or less approximating the effect of a room receiving daylight through its windows. Or, throwing aside any purpose to simulate daylight in intensity or distribution, he can put artificial light simply where it is needed, merely furnishing such a groundwork of general illumination as will serve the ends of art and convenience....'

Bell made it clear where he stood on this issue, advocating strongly the efficient use of lighting by skilful application related to needs. His views accorded with the thinking of the pioneering illuminating engineers of his day, for whom efficiency was a prime motive force. Their innovations of Luminous Intensity, Luminous Flux, Illumination (now termed illuminance) and Luminance evolved from determination to achieve efficiently engineered applications of the new light sources. They remain the basic concepts of photometry, and are dealt with in any textbook of illuminating engineering. However, to gain some feel for the circumstances of their development, the reader is strongly advised to seek out an early edition of Walsh's *Photometry*.[4]

The other approach identified by Bell, that of approximating daylight, did not lack less efficiency-conscious promoters. Banham[5] has reviewed the evolution of lighting practice from the era of the central chandelier, and has identified the innovation of 'architectural lighting' to 'exploit the left spaces and hollows of normal construction to provide light without making its sources visible' with a cross-fertilisation of ideas between architects working in California and in Berlin during the 1920s. Their techniques for achieving luminous panels and strips in their buildings led naturally to the luminous ceiling that first appeared in the late 1930s; and with the innovation of the tubular fluorescent lamp and the availability of translucent plastic sheeting materials, such installations became numerous.

It was in the immediate post-war years that there was a notable attempt in the USA to achieve a major revision of the conventional photometry. Moon and Spencer[6] proposed a novel system of photometry that enabled them to define eight factors of a lighting installation that permit objective evaluation of its suitability for either utilitarian or decorative interiors. Compared with the conventional photometry, the scope of their system was remarkable, providing predictable measures relating to the relative visibility of a task and to the formations of shadows and specular reflections. Some factors, such as aesthetic and psychological effects, inevitably had less precise procedures for rating.

When they applied their attentions to the luminous ceiling, it might have seemed that the architectural and the engineering approaches were in perfect accord, for although the luminous ceiling rated only 'good' for psychological effect (on a scale from poor to excellent), in all other respects it was a veritable panacea. Perhaps their enthusiasm stemmed from their well-known studies of overcast sky distributions, for during the early '50s Moon and Spencer published extensively on the optimisation of the performances of luminous ceilings,[7] and, particularly in the USA, many were installed.

However, it seems that the psychological effect was rather less than 'good'. In 1951 they wrote:[8]

'The feeling is often expressed that ceiling lighting will give an uninteresting, flat effect; but both experiment and theory show this fear to be unfounded.'

However these brave words could not stop the demise of this form of lighting.

The influence of this era upon the development of practice was considerable, particularly in working environments. By comparison, the workspace where the bare lighting needs for adequate vision were applied efficiently and economically to specific task areas appeared gloomy and old-fashioned, and even though the newer lighting techniques that replaced the luminous ceilings did occupy less of the ceiling area, the distribution of illuminance at the working plane usually was similar, and the intensity rather higher.

THE IDEAL OF PERFORMANCE-BASED STANDARDS

In 1933 A. W. Beutell[9] proposed to a meeting of the IES† a possible basis for determining the illuminance required for the performance of a given visual task: this would be the analysis of the task in terms of several visual characteristics that affect the difficulty of performance. Taking as his basis that an easy visual task can be well performed at an illuminance of approximately 10 lx (1·0 ftC), analysis of a more difficult task would indicate a multiplying factor to be applied to this basic illuminance that would compensate for the degree of task difficulty. All that was needed to achieve a scientific basis for the prescription of illuminance was a set of reliable scales of multiplying factors.

Beutell's words inspired H. C. Weston, who engaged upon his classical visual performance studies.[10] Using Landolt ring tests to represent tasks of different degrees of difficulty, he established relationships between illuminance over a range of approximately 5–5500 lx and a measure of visual performance that took account of both speed and accuracy. His findings did not altogether confirm Beutell's prophesies. In his own words:[11]

† Throughout this chapter, the initials IES refer, unless indicated otherwise, to the Illuminating Engineering Society of Great Britain, now the Lighting Division of the Chartered Institution of Building Services.

'The only way of making a very poor contrast as visible as a very good one is to view the latter in a visually handicapping illumination, that is, an illumination low enough to depress the contrast sensitivity of the eye. Such a procedure would be irrational.'

Weston resolved the matter by reducing the data to Relative Visual Performance, for which for each type of task the performance achieved at the highest illuminance he examined represented 100%. Figure 1 shows plots of 90% RVP derived from Weston's data[10] and these may be compared with the IES Illumination Scales that appeared in the 1955 IES Code. For six categories of task size, the appropriate illuminance is selected from one of three scales. The correspondence of the A scale to Weston's results for high contrast tasks is apparent, but the scope of the B and C scales, which are to take account not only of lower contrast but also lower task lightness, is limited as Weston deemed that to keep illuminances reasonable, the extent of this range should be limited arbitrarily to a factor of 10. The influence of size was not limited in this way and covers a 100-fold range of illuminance, and is, as can be seen, extrapolated some way beyond the limit of the experimental data. However, the workings of the system seemed to be simple enough: general office work is classified as a 4B task, and the corresponding illuminance is 200 lx.

Nevertheless, the practitioners demanded more simplification. In particular, the need to define task contrast was an excessive complication, and so in 1961 Weston[11] proposed a simplified method of determining illuminance that depends upon only the size of detail and the lightness of the task. The heavy dotted line in Fig. 1 shows illuminance by this procedure when reflectance is 0·9, which was the background reflectance for the experimental tasks. It can be seen that not only does this procedure ignore task contrast, but it diminishes the influence of size of detail. In this way, the relationships established by the performance studies hardly appear in the procedure for determining task illuminances.

This procedure provided the basis for the revised schedule of illuminance in the 1961 IES Code.

Meanwhile the American IES had adopted a procedure based on Blackwell's proposals. By measuring a real task with a Visual Task Evaluator, this procedure enables the determination of the illuminance required to provide a standard of performance defined as 99% accuracy of detection at 5 assimilations per second: this standard was adopted by the American IES as the basis of their illuminance recommendations.

The Fourth Edition of the American IES Handbook[12] quotes the results

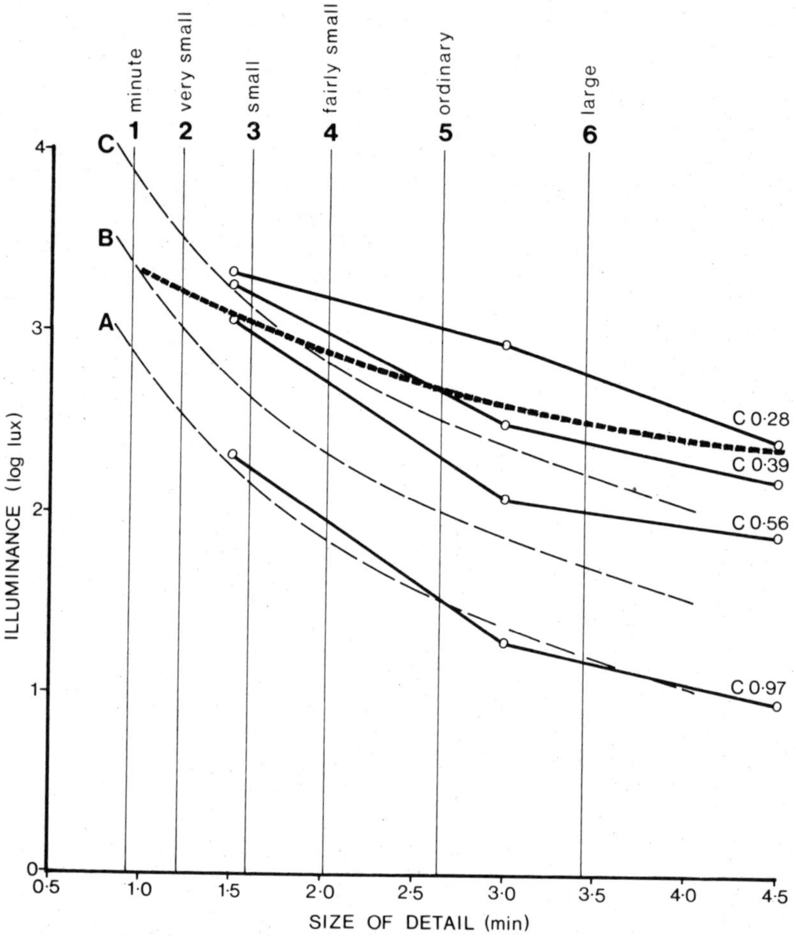

FIG. 1. Illuminance standards developed from Weston's visual performance studies. —O— 90% relative visual performance from Weston's data[10] for different values of contrast (c). — — IES Illuminance Scales (1955). Scales B and C are for medium or low task lightness and/or contrast. ▪▪▪▪ Basic Task Illuminance for reflectance 0·9, from Weston's 1961 proposal $L_b = 180\,S^{-1\cdot5}$ ftL.

of a series of field measurements. In a school, examples of black print on white paper were studied. For 10-point type, as is used for the text of this book, an illuminance of 10 lx is required to provide for the standard visual performance.

For 8-point type, the illuminance required increases to 12 lx,

and 6-point type demands 32 lx, but on the other hand,

12-point type needs only 6 lx.

In an office, a sample of good quality typewritten material was found to require 11 lx.

It seems, then, not unreasonable to ask why the same handbook recommends 323 lx for reading printed material in schools, and 1100 lx for regular office work.

In the same surveys, samples of spirit duplicated material were studied in the school. The quality was variable; the median requirement was 1380 lx, and for poorer quality copies, 7360 lx. In the office, a typed fifth carbon copy required 1430 lx, and a typed original made with an 'extremely poor ribbon', an astronomical 33 800 lx. With such massive variations of illuminance requirement among differences of visual task that may be commonly encountered in a relatively straightforward work occupation, the prescription of a recommended illuminance by this procedure depends upon the precise nature of the visual task that is selected as representing the visual content of the work. In the end, the recommended standard must depend more upon human judgement than upon scientific measurement.

Weston was sharply critical of Blackwell's approach of defining a single performance standard to be achieved for all classes of visual task,[11] and his words quoted earlier in this chapter must come to mind, but finally he was unable to meet the demands of practitioners for a usable system of prescribing performance-based standards of illuminance. At the time of writing, the 1961 IES Code has been superseded three times, involving substantial increases of recommended illuminances made without reference to visual performance.

Beutell's notion had been that, armed with scientific knowledge, the lighting engineer would go out into the world analysing tasks and prescribing lighting to meet visual needs. Beutell had not foreseen that the more obvious use of this new knowledge would be to make visual tasks more simple, rather than to make lighting more complicated. The difficult tasks quoted from the American surveys have been eliminated from all but the most archaic workplaces by modern typewriters, duplicators and business machines. There are many instances from industry of visually difficult operations having been simplified or eliminated by modern control systems, and in fact the increasingly widespread use of cathode ray tube and light emitting diode displays could support the argument that for general lighting, high illuminances should be avoided in the interests of task visibility. For the most part, elaborate task lighting is now restricted to situations where it is impossible either to re-design the task or to automate the operation, as in surgery.

THE IDEAL OF THE DESIGN OF THE VISUAL FIELD

In 1953 the National Illumination Committee of Great Britain received a report from its sub-committee on Lighting Principles entitled 'The Design of the Visual Field'.[13] This report attempted to step beyond the limitations of the Lumen Method that dominated design practice at the time, and to review the gamut of components that constitute the visual experience of an illuminated space. Lighting was considered in relation to architecture, decoration, furnishing, emphasis, modelling, comfort and efficiency.

The correctness of the report's observations was readily acknowledged, and to this day it remains an outstanding statement of the principles of interior lighting. What was needed to achieve a major advance in the general quality of design practice was a design method capable of co-ordinating these components, so enabling the purposeful application of light in the knowledge of its eventual effect. The obvious contender, in fact the only contender for the purpose, was the Designed Appearance Method of J. M. Waldram,[14] who had been the chairman of the sub-committee that had produced the report.

As this method is the subject of Chapter 5 of this book it will not be described here apart from noting that its objective is the control of the appearance of a room and its contents by the manipulation of lighting. The designer specifies his intentions by according a value of apparent brightness to each element of the field of view; he uses apparent brightness curves to convert these values to luminances, and he uses illuminating engineering techniques to develop a lighting specification to achieve these luminances.

The report had aroused much interest and discussion, but during the ensuing decade there was no appreciable sign that its proposals were being adopted by general lighting design practice. In 1965 Hopkinson[15] proposed a way forward. His objective was a Brightness Code: he saw the 1961 IES Code with its Glare Index procedure to have been a step in the right direction, and the next step would be a code that would specify design objectives in terms of luminance.

It is necessary for the purpose of this chapter to examine why it was that these events did not come to pass. It is easy to blame it all on to indifference, but that would conceal some considered doubts and objections.

It was Hopkinson who had provided the original scales of apparent brightness[16] from which Waldram had developed his method, but he had later expressed some reservations regarding this approach to design that were due to the phenomenon of lightness constancy, or as it is sometimes called, brightness constancy.[17] This may be described as the perceptual ability to distinguish light and dark coloured surfaces largely independently of their lighting, and clearly such an ability must raise some doubts concerning a procedure based on the premise that the appearance of a surface is determined primarily by its luminance.

In 1965, Jay[18] expressed his reservations. He argued that under normal conditions of interior lighting, an illuminated surface is perceived as having a certain lightness and a certain illuminance, and that whereas a given increase of luminance may be achieved alternatively by an appropriate increase of lightness or of illuminance, perceptually these alternatives are not equivalents. Consequently, the very notion of apparent brightness was of doubtful validity, and designers would be better advised to concern themselves with the separate control of two factors that are readily understood: lightness (or reflectance) and illuminance.

Lynes[19] has examined the circumstances that encourage the perceptual constancies to be maintained and has concluded that the visual conditions achieved by the diffuse, uniform, limited glare lighting that is provided in most of our building interiors effectively ensures that lightness constancy will hold. The situation changes when the visual field is composed of strong contrasts and particularly when the sources of illumination are not visible, as in display lighting or dramatic stage lighting, when constancy may break down completely. Under these circumstances, changes of lightness and illuminance may be perceptually equivalent.

Waldram has demonstrated his applications of the designed appearance method with some superbly engineered installations, the best known being his lighting schemes for some British cathedrals.[20] These clearly have been situations in which, whether with the intention or not, break-downs of constancy have been engineered. It is suggested here that it was not correct to suppose that procedures developed to achieve the engineering of criteria such as emphasis and modelling could be as successfully or as purposefully applied to engineering the appearances of surfaces where there is no obvious requirement for emphasis, and the distribution of lighting is planned principally with regard to the avoidance of excessive illuminance diversity, and of high brightnesses and dense shadows.

THE QUEST FOR QUALITY

The sum total of the foregoing review of three mainstreams of thought has every appearance of a dismal tale of failure and frustration. The lighting profession had developed a formidable technology, but lacked objectives. It looked outwards, seeking to involve other professions. Appraisal sessions were organised,[21] and questionnaires were circulated in efforts to identify what are the physical aspects of the visual field that contribute towards quality, and even, what is quality?

The author is of the opinion that a reasonably coherent outcome of this period of soul-searching is becoming recognisable. This has not been achieved by the discovery of new or hitherto unidentified objectives, but has grown from a realisation that the fundamental thinking behind the three already reported and apparently rejected mainstreams is basically sound.

Apparently new ideas that have fitted this pattern have tended to be well received by a profession that has been anxious to demonstrate that really it always knew that it was something like that.

Accordingly, the remainder of this chapter comprises re-examinations of the three mainstreams.

COMING TO TERMS WITH INDOOR DAYLIGHT

In temperate climates, an overcast sky is a standard outdoor condition for daylight design, but as Lynes has made clear,[19] this is far from being an ideal distribution of light for revealing surface colours and textures. The attractive features of daylight become evident outdoors when the sun breaks through the clouds, or indoors when the source is a mere patch of sky framed by a window. Efforts to bring the overcast sky indoors were bound to lead to rejection.

Reporting an extensive literature survey of psychological reactions to environments with and without windows, Collins concluded:[22]

'Although there are still questions about the physiological requirements for daylight and sunshine, the importance of these for the psychological well-being of a building's inhabitants should not be underestimated. Daylight and sunshine contribute a highly dynamic character to the internal environment of a room. They introduce change and variety into an otherwise static atmosphere. The addition of dynamic qualities through the provision of view, daylight and sunshine appears to be one of the most vital functions performed by a window.'

This emphasis upon dynamic qualities, with its suggestions of force, conveys a notion that relates readily to experience. When comparing Figs. 2 and 3 it is necessary to remind oneself that apart from the girl's dress, the only change is a change of lighting, and yet the whole experience of the space and the objects within it has been affected.

In 1966, Lynes and a group of colleagues from the Pilkington Daylight Advisory Service presented a paper to the IES entitled 'The Flow of Light into Buildings'.[23] In this paper the experience of the team had been related by Lynes to photometric concepts first propounded by Gershun 30 years earlier.[24] Illuminance at a point in space may be measured both as a scalar quantity, where scalar illuminance (E_s) is equal to the average illuminance over the whole surface of an infinitesimal sphere located at the point, and as a vector quantity. The difference of illuminance on opposite sides of a plane passing through the point due to a single point source varies in accordance with the Cosine Law as the plane is rotated with respect to the source, so that a rotation of 360° is represented as a sine wave. Addition of a second source, or any number of subsequent sources, always resolves to a sine wave with the same frequency. Its amplitude is taken to be positive in the direction of the higher illuminance, and this amplitude defines the magnitude of the illumination vector ($\vec{\mathbf{E}}$) and the direction in which it occurs.[23]

The total distribution of illuminance about a point can be represented as an illumination solid,[25] the average value of which is equal to the scalar illuminance at the point. Figure 4 gives a section through such a solid, which can always be shown to be the sum of two component solids: the vector solid is a sphere tangential to the point that is defined completely by the magnitude and the direction of the illumination vector, and the remainder is the symmetrical solid, which is always perfectly symmetrical about the reference point. The vector/scalar ratio ($\vec{\mathbf{E}}/E_s$) is a convenient measure of the asymmetry of the illumination solid, but it should be noted that it does not define the illumination solid as the shape of the symmetrical solid is not specified.

The vector/scalar ratio and the vector direction relate to the perceived strength and direction of the 'flow of light'. The understanding of this term has caused some difficulty.

FIG. 2. Entrance hall to an office building by day. (Courtesy Pilkington Brothers Ltd.)

FIG. 3. Entrance hall to an office building by night. (Courtesy Pilkington Brothers Ltd.)

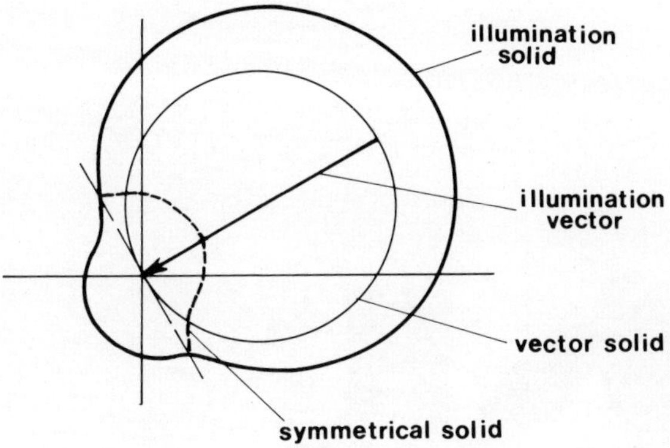

FIG. 4. Section through an illuminance solid, showing component solids.

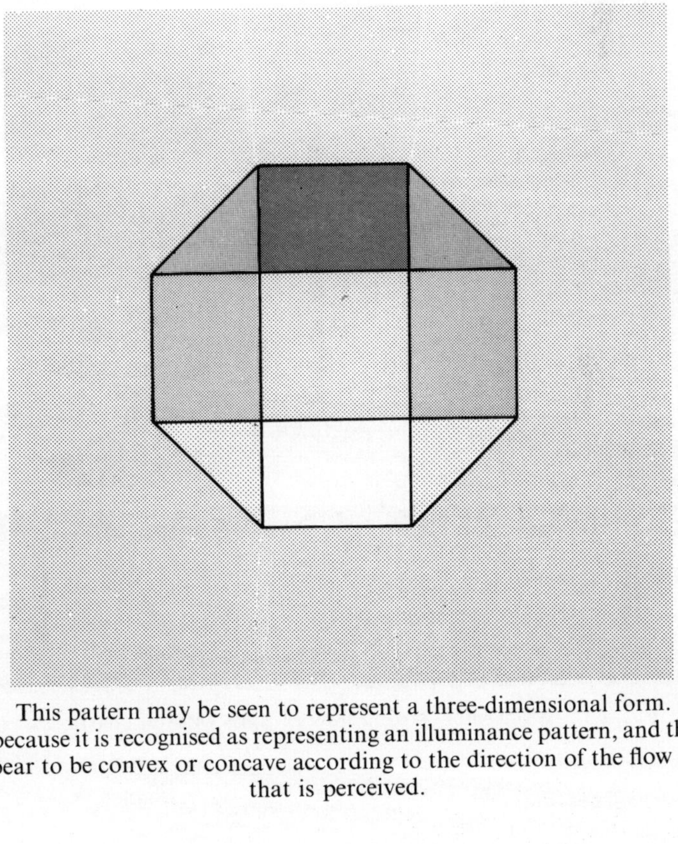

FIG. 5. This pattern may be seen to represent a three-dimensional form. If so, it will be because it is recognised as representing an illuminance pattern, and the form will appear to be convex or concave according to the direction of the flow of light that is perceived.

FIG. 6a. Moderately strong flow of light without sharpness.

FIG. 6b. Moderately strong flow of light with sharpness.

Figure 5 shows a simple geometric pattern that readily gives rise to perceptions of three-dimensional forms, the perceived form alternating between convex and concave according to which way up the page is held. Although most readers will be familiar with illusions of this sort, the figure is reproduced to enable experimentation. A sense of a directional flow of light is a prerequisite for this figure to promote perception of depth, and common experience favours the assumption that light descends from above. This expectation can be overridden. In the vicinity of a window the figure works better on its side, or even placed flat on a table and rotated to achieve reversals. The presence of the window and the patterns of light and shade that it imparts to all of the objects and the surfaces that make up the environment enforce a sense of flow of light that is more powerful than the normal expectation.

Figure 6a shows three objects that have responded in distinctly different ways to a condition of directional lighting. The earthenware pot shows a clear 'illuminance pattern', the appearance of the globe is dominated by a 'highlight pattern', and the lampshade casts an intricate 'shadow pattern'. Although distinctly different, these appearances are utterly consistent with a single overall notion of a flow of light. In Fig. 6b the appearances are different, but it is argued that this is not a difference of either the strength or the direction of the flow of light; Cuttle has used the term 'sharpness' to describe this difference.[26] The clean-cut appearances of the highlight pattern and the shadow pattern indicate a compact source, but the appearances of illuminance patterns on the pot in the two cases are virtually identical. For these two photographs, the lighting was adjusted to achieve matching vector/scalar ratios and vector directions.

Every opaque, solid object in directional lighting forms an illuminance pattern, and it is the appearances of these illuminance patterns that promote the perception of a flow of light within a room. Vector/scalar ratio and vector direction are photometric concepts that correspond well with this subjective evaluation. In addition, the physical characteristics of the light sources determine the sharpness of the highlight and shadow patterns. These will be evident on surfaces that have appropriate properties, and add a separate aspect of appearance due to directional lighting.

Cuttle[26] has suggested that understanding the roles of these three lighting patterns is crucial to appreciating the interaction of lighting and object appearance. For general lighting purposes, the objective should be not to achieve a particular type of object appearance, but to provide lighting that has the capability of producing variety of object appearance. Attention is turned from considering what luminance pattern might be produced on a particular object, to considering forms of lighting that create separately identifiable illuminance, highlight, and shadow patterns, for this will be lighting that makes matt surfaces appear distinct from glossy surfaces, and convex surfaces distinct from concave ones. Obviously a very weak flow of light will be inappropriate, but if it is too strong the three lighting patterns merge and their separate identities become lost. (This could be the point at which apparent brightness starts to become reliable.) For general lighting, moderation is called for, both of the flow and of the sharpness of light.

COMING TO TERMS WITH PERFORMANCE-BASED STANDARDS

Recognition of the benefits that accrue from professional integration during the design of a building and its services has exposed the basis of illuminating engineering to wider scrutiny, and some building services engineers have expressed surprise upon discovering that there is no recognised procedure that provides a reasonably precise specification of lighting needs for optimum human performance or comfort. As understanding spreads of the highly influential role of lighting, whether natural or artificial, as it affects the energy-balance of a building and the control of its indoor environment, the lighting profession is under increasing pressure to justify the standards of lighting that it recommends. There is a heartfelt need for a measure of lighting that relates to its value, or to its usefulness.

In 1972 the Commission Internationale de l'Eclairage published a technical committee report[27] in which it was proposed that for working situations the lighting design criterion should be Visual Performance Potential, where this criterion is defined by an appropriate measure of task visibility.

In the same year, the American IES published the 5th edition of its Handbook in which illuminance standards for many working situations are specified in terms of Equivalent Sphere Illuminance (ESI). In practical terms, this really is not an illuminance at all, but is a specification of relative task visibility which can be understood as follows.

The basic visual data upon which the proposed CIE system is based derives from Blackwell's experimental studies, and is specified for Reference Illumination, this being a precisely defined standard of lighting.[28] It is approximated by a photometric sphere illuminated by a filament source, and so it would be possible (even if not practical) to measure the visibility of a task in a workplace, and then to place it in a sphere and view it through a visibility meter, reproducing the direction of view in the workplace. The sphere illuminance that achieves the same task visibility as was measured in the workplace is the ESI of the task in the workplace.

Various models of Visual Task Evaluators have been made to Blackwell's specifications[29] which are precision contrast reducing visibility meters that provide facility for the operator to follow his setting for task visibility threshold by adjusting the contrast of a standard task for threshold at the same field luminance. This provides a measure of Equivalent Contrast by which the task being studied can be related to basic visual data, enabling, among other things, the ESI to be computed. However, equipment of this sort is seldom found outside specialist lighting laboratories.

Visibility meters that enable comparisons of the visibilities of different tasks, or the same task under different forms of lighting, are relatively easy to construct and to operate, and recently the technical literature has carried descriptions of a number of ingenious instruments for this purpose. Clearly, a simple instrument capable of giving acceptably reliable measures of relative task visibility could provide measures of ESI if the task could be examined under sphere lighting as well as under the workplace lighting. Ngai *et al.*[30] have described an ESI meter that embodies sphere lighting to enable successive settings, and Slater[31] has described the use of a standard task taken to be representative of office reading tasks, and calibrated with a simple visibility meter under sphere lighting.

The measured value of ESI can be appreciably different from the task illuminance (E). Several factors can account for this, and proposals for a comprehensive system of evaluation of these factors is awaited from the CIE committee concerned, but the crucial influence of veiling reflections upon the visibilities of common tasks is now widely recognised. The Contrast Rendering Factor (CRF) is a measure of the visibility of a task in the workplace relative to the visibility that would be achieved by reference lighting at the same task luminance.[27]

A more convenient and readily understandable index is the Lighting Effectiveness Factor (LEF), which is simply the ratio of ESI to E. Like CRF, it indicates the effectiveness with which lighting is applied for the purpose of revealing the detail of the task, but the scale of values has practical relevance as it indicates the extent to which increased lighting of the same quality is required (or reduced illuminance would be permitted) to achieve a certain standard of relative task visibility.

As has been mentioned, the effects of veiling reflections can be crucial, and it is by no means uncommon for surveys of conventional lighting installations to reveal LEFs of less than 0·5. However, reference lighting is by no means an optimum condition, and by carefully relating the directions of incident light in relation to the direction of view and the reflection characteristics of the task, and in some cases, the state of polarisation of the light, LEFs in excess of 2·0 have been achieved.

The energy implications of this form of economy are considerable. Cuttle and Slater have proposed the Visibility Efficacy[32] to indicate the extent to which this is achieved. The ESI at a workstation is multiplied by the portion of the total workplace area that is attributable to that workstation, and is divided by the electrical power rating of the lighting of that area to give the visibility efficacy in equivalent sphere lumens per watt.

While it cannot be denied that so far practical experience of applications of these new measures of lighting is fairly limited, at least outside North America, interest has been growing rapidly. Much effort is being devoted towards the development of a recognised procedure for predicting ESI and LEF, and novel measuring instruments have been the subject of considerable ingenuity. It is suggested that this level of activity is promoted by the expectation that general acceptance of these measures will go a long way towards meeting the profession's need for measures that reflect the benefits and worth of lighting.

COMING TO TERMS WITH THE DESIGN OF THE VISUAL FIELD

The term Luminance does not appear in any of the recognised schedules of standards for indoor lighting, but this should not be taken to mean that the notions that the advocates of luminance had in mind have been rejected.

More than 50 years have passed since it was pointed out that illuminance can be used for more than the simple purpose of specifying the density of lumens incident upon the horizontal working plane. In 1925 Luckiesh and Holladay[33] in their classic study of disability glare specified the weighted average luminance of the field of view in terms of illuminance at the eye normal to the direction of view. This is a use of the concept of illuminance that is quite different from its conventional applications, and in fact much of the recent discussion of the relative merits of different illuminance concepts has been complicated by confusion of this distinction.

Measures of illuminance at a point are distinguished by the receiving surface that is specified, each surface specification defining a certain weighting to be applied to the luminance field surrounding that point. Some of these surface specifications, such as the horizontal working plane, mean spherical or scalar illuminance, and the illumination vector, have been developed with the objective of providing some yardstick for overall comparisons of how the appearances of surfaces and objects are affected by the light that is incident upon them from their surrounding luminance fields. Others have been developed to provide bases for comparisons of the overall response to a visual field due to the light from it reaching the eye. Illuminance at the eye is of this latter origin, and it has been used by Hopkinson[17] and by Helson[34] to specify adaptation level for the purpose of determining apparent brightnesses. Waldram uses a more restricted cone.[14]

Mean cylindrical illuminance (E_c), being the average illuminance over the curved surface of a vertical infinitesimal cylinder, represents the average illuminance at the eye for a person who maintains a horizontal line of sight as he looks around him. This concept has been used for specifying illuminance standards in the USSR[35] for non-working locations following studies by Epaneshnikov and Sidorova[36] that showed that assessments of the adequacy of lighting of a number of public buildings correlated better with measurements of mean cylindrical illuminance than of illuminance on a horizontal plane.

Mean cylindrical illuminance certainly appears to provide a more plausible model of human behaviour than does scalar illuminance when applied for this purpose, for as Lynes has expressed it, 'the scalar illuminance at a given point may be considered equal to the average adaptation luminance when you stand at that point and move your head around at random'.[37] However, it appears that for regular overhead lighting installations, a fairly high degree of correspondence between average scalar and average mean cylindrical illuminances can be expected, in that a change in the lighting installation that alters one illuminance is likely to produce a corresponding change of the other. Lynes has used his Luminaire Domain theory[38] to examine this correspondence, but had to reserve judgement as it appears that neither of these measures is predictable with sufficient precision by the BZ system.

The current IES code specifies lighting standards for some non-working locations in terms of scalar illuminance, which presumably recognises its wider usefulness (as has been discussed) rather than any superiority over mean cylindrical illuminance as an index of the adequacy of the lighting of a room. It should be noted that for non-regular lighting, such as daylight from a side window, correspondence of scalar and mean cylindrical illuminances can not be expected, but these are situations for which neither a single measure nor an average of measures is likely to provide a very satisfactory index of lighting adequacy.

The foregoing concerns attempts to integrate the effect of the distribution of luminances contained in the visual field, but as has been discussed, a procedure of designing lighting to achieve a prescribed distribution of luminances is not necessarily appropriate to locations where contrasts are to be low. It is not uncommon for practitioners who engage in the more visually exciting types of lighting design work to implore their pedestrian brothers to learn from their own repertoires of techniques, but the bulk of lighting practice is directed by objectives to which most of these techniques are inappropriate. The achievement of good visual conditions for work is unlikely to be compatible with the achievement of a visually exciting environment, and calls for recognition of different sets of objectives and procedures.

For working locations, or in fact any location where uniformity of illuminance and visual comfort are more important criteria than emphasis or modelling, it remains the conventional practice to consider the distributions of reflectance and illuminance as two different but related factors.

The IES Code recommends that surfaces should be as light as is practical, both to diffuse light by reflection, and in the cases of the ceiling and window walls, to prevent light sources being seen against dark surrounds.

The ceiling is to be finished the ubiquitous white to achieve a reflectance of not less than 0·7, and the minimum recommended reflectances for the walls and floor are 0·3 and 0·2 respectively. Following a study by Jay,[39] who considered a range of lighting criteria in rooms where the foregoing conditions apply, the code recommends that the wall illuminance ratio (ratio of average wall illuminance to average illuminance of horizontal working plane) is to be in range 0·5–0·8, and the ceiling illuminance ratio in the range 0·3–0·9. There has been criticism of these recommendations that they are recipes for blandness, but observation indicates that what is generally recognised to be good current practice for the lighting of working locations demonstrates compliance, and so unless a designer has some good reason to re-order his priorities, he should be advised to adhere to them.

The BZ system[40] dates back to 1961, and it has been presumed that readers are already familiar with its workings. The distributions of light outputs of luminaires are classified by a BZ class number and a value of Flux Fraction Ratio, the system having been proposed originally as a means of enabling utilisation factors to be computed and so dispensing with the need for separate tables of utilisation factors for each type of luminaire. In fact, luminaire manufacturers have continued to publish their separate tables and these have proved more popular than the BZ utilance data, but the achievement of the system has been to provide a fairly simple means of predicting a Glare Index[41] that relates to the direct discomfort glare due to a regular installation of luminaires.

More recently the full usefulness of the BZ system has become recognised. To state a BZ number and flux fraction ratio at a given room index is, in fact, to specify the relative direct illuminances of the ceiling, walls and working plane. Coomber and Jay[42] showed how the use of the system could be extended to provide a simple means of predicting the average luminances of these surfaces allowing for interreflections, and Cuttle[43] extended the arguments of Jay's previously mentioned study[39] to propose a design method that copes simultaneously with several design criteria.

This latter approach has been developed to produce the Multiple Criterion Design Method.[44] For a situation in which the IES recommendation that the Uniformity Ratio (ratio of minimum to average illuminance of working plane) should be not less than 0·8, the designer specifies several criteria that he wishes to achieve. These may include the average illuminance of the working plane or the average scalar illuminance at eye level, the limiting glare index, wall and ceiling illuminance ratios and the average vector/scalar ratio. The method enables the designer to modify some of these criteria if the combination that he has specified is unattainable, until he has defined the BZ number, flux fraction ratio, lumen output and minimum luminous area of a luminaire that will simultaneously satisfy the attainable combination of his selected criteria.

In this way, quite small differences between the photometric performances of different luminaires may be examined in terms of their overall suitability for providing lighting that satisfies a fairly broad range of design criteria.

CONCLUDING REMARKS

To stand back from recent developments and to set them into a perspective is not easy, even with the author's geographical advantages, and no doubt time will reveal misplacements.

It has been the thesis of this chapter that development has revolved around three fairly basic objectives which have found different forms of expression over the years. It is in this context that the innovation that has enriched our technology has occurred, and the author is pleased to record that he foresees no imminent conclusion of this process.

REFERENCES

1. CIE Publication No. 17 (E-1.1) 'International Lighting Vocabulary', 1970, Commission Internationale de l'Eclairage, Paris.
2. O'Dea, W. T., *The Social History of Lighting*, 1958, Routledge and Kegan Paul, London, p. 60.
3. Bell, L., *The Art of Illumination*, 2nd ed., 1912, New York, p. 208.
4. Walsh, J. W. T., *Photometry*, 1st ed., 1926, 2nd ed., 1953. Constable, London.
5. Banham, R., *The Architecture of the Well-Tempered Environment*, 1969, Architectural Press, London.
6. Moon, P. and Spencer, D. E., *Lighting Design*, 1948, Addison Wesley, Cambridge, Mass.
7. Moon, P. and Spencer, D. E., *Illum. Engng* (*New York*), 1949, **44,** 465 and *ibid* 1951, **46.**
8. Moon, P. and Spencer, D. E., *J. Franklin Inst*., 1951, **251,** 453.
9. Beutell, A. W., *Illum. Engr.* (*London*), 1934, **27,** 5.
10. Weston, H. C., 'The Relation between Illumination and Visual Performance', Industrial Health Research Board Report No. 87, 1945 (reprinted 1953), HMSO, London.
11. Weston, H. C., *Trans. Illum. Engng Soc.* (*London*), 1961, **26,** 1.
12. *IES Lighting Handbook*, 4th ed., 1966, Illum. Engng. Soc., New York, p 2–17.
13. 'The Design of the Visual Field', Report of sub-committee of the National Illumination Committee of Great Britain, *Trans. Illum. Engng. Soc.* (*London*), 1953, **18,** 225.
14. Waldram, J. M., *Trans. Illum. Engng. Soc.* (*London*), 1954, **19,** 95.
15. Hopkinson, R. G., *Trans. Illum. Engng. Soc.* (*London*), 1965, **30,** 63.
16. Hopkinson, R. G., Stevens, W. R. and Waldram, J. M., *Trans. Illum. Engng. Soc.* (*London*), 1941, **6,** 39.
17. Hopkinson, R. G., Proc. CIE (Brussels 1959), 1960, Commission Internationale de l'Eclairage, Paris; reprinted in Hopkinson, R. G., *Architectural Physics: Lighting*, 1963, HMSO, London.
18. Jay, P. A., *Light and Ltg*, 1967, **60,** 42.
19. Lynes, J. A., *Ltg Res. Technol.*, 1971, **3,** 24.
20. Waldram, J. M., *Trans. Illum. Engng Soc.* (*London*), 1959, **24,** 42.
21. Hewitt, H., Bridgers, D. J. and Simons, R. H., *Trans. Illum. Engng. Soc.* (*London*), 1965, **30,** 91.
22. Collins, B. L., 'Windows and People: A literature survey', NBS Building Science Series 70, 1975, US Dept. of Commerce and National Bureau of Standards, Washington.
23. Lynes, J. A., Jackson, G. K., Burt, W. and Cuttle, C., *Trans. Illum. Engng Soc.* (*London*), 1966, **31,** 65.

24. GERSHUN, A. A., 'Svetovoe Pole,' 1936, Moscow; English translation by Moon, P. and Timoshenko, G., *J. Maths and Phys.*, 1939, **18,** 51.
25. CUTTLE, C., *Ltg Res. Technol.*, 1971, **3,** 171.
26. CUTTLE, C., in *Architectural Psychology: the Proceedings of the Lund Conference*, (ed. R. Kuller,) 1973, Studentlitteratur, Lund, Sweden, p. 12–22.
27. CIE Publication No. 19 (TC-3.1), 'A Unified Framework of Methods for evaluating Visual Performance aspects of Lighting', 1972, Commission Internationale de l'Eclairage, Paris.
28. 'Implementation Procedures for evaluating Visual Performance aspects of Lighting', preliminary 4th draft, 1975, to be published by Commission Internationale de l'Eclairage, Paris.
29. BLACKWELL, H. R., *Illum. Engng* (*New York*), 1970, **65,** 267.
30. NGAI, P. Y., ZELLER, R. D. and GRIFFITH, J. W., *Ltg. Des. & Applic.*, 1975, **5,** 32.
31. SLATER, A. I., *Ltg Res. Technol.*, 1975, **7,** 52.
32. CUTTLE, C. and SLATER, A. I., Proc. CIE (London 1975), 1976, Commission Internationale de l'Eclairage, Paris.
33. LUCKIESH, M. and HOLLADAY, L. L., *Trans. Illum. Engng. Soc.* (*New York*), 1925, **20,** 221.
34. HELSON, H., *Adaptation-Level Theory*, 1964, Harper & Row, New York.
35. Building Standards & Specification (SNip), Section II-A, 9-71 (1971) (in Russian).
36. EPANESHNIKOV, M. M. and SIDOROVA, T. N., *Svetotekhnika*, 1965, 11.
37. LYNES, J. A., in discussion ref. 15.
38. LYNES, J. A., *Ltg Res. Technol.*, 1975, **7,** 185 and 242.
39. Jay, P. A., *Trans. Illum. Engng Soc.* (*London*), 1968, **33,** 47.
40. 'The Calculation of Utilisation Factors—the BZ Method,' IES Tech. Rpt No. 2, 2nd ed., 1971, Illuminating Engineering Society, London.
41. 'Evaluation of Discomfort Glare—the IES Glare Index System for Lighting Installations', IES Tech. Rpt. No. 10, 1967, Illuminating Engineering Society, London.
42. COOMBER, D. C. and JAY, P. A., 'A Simplified Method of Calculation for Luminance Ratio and Designed Appearance Lighting Installations', IES Monograph No. 10, 1967, Illuminating Engineering Society, London.
43. CUTTLE, C., *Architects' J.*, 1969, **150,** 245.
44. 'Multiple Criterion Design: a Design Method for Interior Electric Lighting Installations', IES Tech. Rpt. No. 15, 1977, Illuminating Engineering Society, London.

Reference is made also to the Codes for Interior Lighting published under various titles in 1955, 1961, 1968, 1973 and 1977 by the Illuminating Engineering Society, London, (now the Chartered Institution of Building Services).

Chapter 2

LIGHT AND VISION

G. T. YONEMURA
Ph.D.

Research Psychologist, Institute of Applied Technology,
National Bureau of Standards, Washington, USA

SUMMARY

Three levels of visual information needs are defined. The psychological responses associated with these information requirements are identified and correlated with their stimulus counterparts. The criteria for recommending levels of illumination should be 'goodness of seeing' rather than 'just barely able to see'. Blur is introduced as a parameter that has been neglected in task description for lighting. The Modulation Transfer Function is recommended as a technique that analytically describes the classical parameters associated with visual sensory performance as well as blur. The relative contributions of physiological and psychological variables in visual task performance studies are discussed.

What is vision? Is our understanding of the visual processes, as we know them today, sufficient for application to lighting? Are the classical techniques used in investigating the visual processes appropriate to the problem of lighting the practical working environment? These questions must be addressed and resolved before a valid and non-controversial basis for recommending levels of illuminance for everyday visual task performance can be realised.

In extracting information from the visual field, the psychophysical performance required of the observer depends on the type of visual information required. Visual tasks differ significantly in difficulty and complexity. At the simplest level is detection—determining the presence or absence of an object in the field of view. In the laboratory we can have observers detect the presence of a disc. In industry, a practical example is the detection of a crack in machined or welded parts as seen on an X-radiograph. The defect is seen as an inhomogeneity in an otherwise homogeneous field and can range in visibility from near threshold (just barely discernible hairline inhomogeneity) to highly visible.

The next higher order of visual task performance requires the observer not only to detect the presence of an object, but also to recognise it as a member of a group. This level will generally require shape or form discrimination. Laboratory recognition tasks will be the discrimination of geometric forms or the determination of the minimum separation between two parallel lines. Applications to the real world may be of the form; 'The object I see is a triangle, a man, a letter, etc.'

Detection and recognition are followed by identification. This level requires prior performance of the first and second levels, and in addition the observer must be able to discriminate between members of the same group. The additional performance requirement is the perception of details, for example, facial features, type of letter, etc. A simple paradigm of the above discussion is: 'I see something,' (detection), 'it is a man' (recognition) 'and his name is John Doe' (identification). Or, 'I see something on the paper, they are letters and the letters are Es'.

Three levels of increasing visual task requirements have been described and differentiated according to the use to which the visual input is applied. A description of the three levels in terms of their psychophysical correlates is necessary to indicate the change in the response counterparts associated with each level. The most basic and simplest response of the visual mechanism is the discrimination of changes in luminance. In detection, this discrimination between two luminance levels, or contrast information, is all that is required. In recognition, 'the task consists of (i) detecting the presence of a signal, then (ii) assigning this signal to a category which has definitive class properties. Whatever theoretical position investigators have adopted, all agree that the contour (outline) of a figure is the "cue" or "information" carrier for shape'.[1] Thus, for detection, a change in luminance sufficient for the determination of the presence of a signal defines the total requirement, but for recognition the additional requirement of contour discrimination is imposed. Identification differs from recognition in that the former involves the perception of details as well as contours and involves aspects of image quality that improve clarity of detail.

All of the above are threshold considerations, just barely permitting performance at some criterion level; e.g. 'I just barely see the spot of light,' (detection), 'I just see that it is a ring,' (recognition), 'I can barely tell that the number is a 3 and not an 8,' (identification). In most everyday tasks involving predominantly a visual component we are not working at threshold levels, or would be unhappy if we had to. For example, a reproduction which is so poor that all we can state is that something is on the paper (detection) is impossible to accept. If the reproduction were slightly better such that we could state that it is printed matter, it would still be intolerable. We could have an improved copy such that we could, with some certainty, identify the letter, i.e. discriminate between an E and B, etc. But just barely being able to discriminate, that is, 'I can barely read the printed page', is still unsatisfactory. The criterion for visual performance should be goodness of seeing, e.g. 'I see the letter E under both conditions, but it is clearer under condition A rather than B'. Just barely being able to detect, recognise or identify an object is unacceptable.

This brings in a parameter of the task, generally ignored in lighting studies, namely blur. The role of blur takes on increasing importance as we proceed through detection and recognition to identification. As pointed out by Perrin,[2] 'the attribute of a photograph that represents the clarity with which details are reproduced may be termed photographic definition, and it is probably determined in large measure by the sharpness of the picture.... If, as seems likely, the most important attribute of definition is sharpness, the criterion of interest must be some property of the microdensitometer trace of the edge of the image—the graph relating the change of density at the edge to the distance across the edge.' Perrin's point of view is supported by Brainard and Caum's[3] results obtained from their studies of differentiation enhancement techniques, which involved modification of the edge gradients by sharpening. They found that the enhancement technique became more effective as the subject's task become more difficult, that is from detection to recognition to identification. For example, by sharpening the edge gradients of their images, the relative improvement in performance was 35% for target detection tasks, but 110% for identification tasks. In another comparison, the relative improvement of 17% obtained for recognition tasks contrasted with a 57% improvement for the corresponding identification tasks. Fox[1] found that recognition thresholds were slightly more affected by edge gradient changes than were detection thresholds. These findings, in agreement with Perrin, indicate the increasing importance of image edge traces as task complexity increases.

The studies discussed above indicate that in the evaluation of image quality for identification, as opposed to simple detection, the edge gradient appears to be an important, if not the critical, factor affecting clarity of detail. In the area of psychophysiology, we have further experimental studies indicating how edge gradients can importantly affect what we see. In his studies of neural interaction, Ratliff found that the sharpness of an image is to a great extent a function of the shape of the edge gradient of the image. Neural interactions accentuate the contrast at sharp spatial gradients and at discontinuities in the retinal image, resulting in a 'crispening' of borders and contours.[4] This phenomenon is called the Mach band effect and has been extensively investigated. A comprehensive review of the experimental studies and neural models to explain this phenomenon has been made by Ratliff.[5] A good example of a similar phenomenon at the physical level is the contrast effect produced by the electrostatic field in the fused dry powder copying process. 'Wherever there is an abrupt change from a light (uncharged) area to a dark (charged) area, more powder is attracted at the edge of the dark area and less at the edge of the light area than in the uniformly dark or uniformly light areas. In order to obtain a copy with continuous tone, a halftone screen is used to break up the image so that there are changes from light to dark everywhere.'[6] As we shall see in the next study to be discussed, not only is the contour between two areas affected by the contrast between those areas, but the contour itself strongly affects the contrast.

O'Brien[7] has demonstrated that, by contour manipulation, the eye can be made to perceive images very different from the physical object distribution. Figure 1 is taken from O'Brien. The three (stimulus) curves to the left show the luminance distributions of the objects. The luminance differences between the left and right sides of each stimulus are above threshold, and in all cases the luminance of the left portion is less than that of the right. The edge gradients are different in all cases. The three (perception) curves to the right of the figure show the brightness distributions perceived when the object luminance distributions are those presented in the curves to the left. At the top we have the typical Mach band. A dark line is perceived on the darker half of the knife edge and a bright line is seen on the lighter half. In the middle pair, although the luminance difference is above threshold, the slowly changing edge gradient results in the perception of a homogeneous field. At the bottom is a dramatic example of how apparent contrast can be reversed by a proper choice of the flux distribution along the edge. Physically the luminance of the left side is lower than the right side, but perceptually the left portion is brighter than the right. Perceived contrast as well as the sharpness of contours are therefore importantly influenced by the shape of the flux distributions along the boundary of an image. Thus, we see that the luminance distribution along

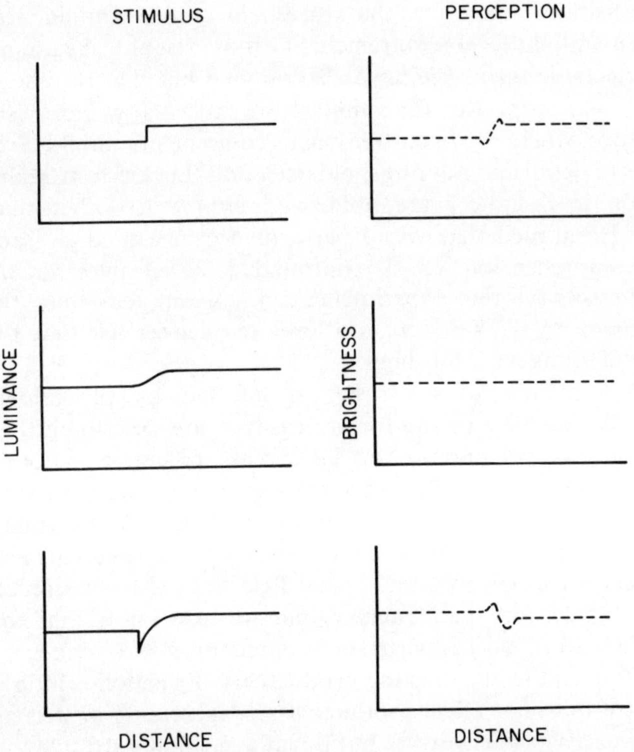

FIG. 1. The effect of edge gradient on perceived contrast. The luminance distribution of the object (stimulus) is given on the left and the perceived image (perception) is shown on the right.

the edge of the detail materially affects the apparent contrast of the target. For this reason a physical measure that is used to describe the clarity of details should have the edge gradient as an important parameter.

The preceding discussions indicate that the psychophysical performance and its physical correlate depend on the type of visual information required. It follows that the most valid studies will be those performed in the real world, that is actual production studies conducted under actual working conditions. This laboratory vs real-world studies controversy had its heyday in the late forties in the US. The two primary antagonists were Luckiesh and Tinker. Luckiesh[8] based his recommendations on threshold acuity experiments, whereas Tinker[9] based his recommendations on task performance under actual conditions using real tasks, e.g. reading comprehension. Tinker felt that threshold studies should not be the determiner of light level requirements as most visual tasks are not of the barely detectable form. We have also pointed out that threshold studies may be inappropriate for application to real-world visual task performance, where the psychophysical requirements should be clarity or sharpness of detail and not threshold detection. Luckiesh strongly felt that production level indices are influenced by factors other than visual capacity. He argued that visual performance measured by production, reading comprehension, etc. is confounded by variables not under the direct control of the experimenter. Luckiesh felt that the levels recommended by Tinker were too low, and Tinker felt that Luckiesh's recommendations were too high.

The performance of a worker is affected by physiological and psychological factors. In the former case we are describing the sensory capacity or process and in the latter case cognitive factors, usually expressed as preferences. The special case we are concerned with in this discussion is where the two processes are dependent on the same physical variable, e.g. luminance. As far as the sensory processes are concerned, an environment in which the total visual field is of the same reflectance is optimum for sensory performance, but we also know that this bland (limited light and shadows) visual environment is not conducive to worker comfort, thus indirectly affecting productivity. Its action is indirect in the sense that it does not affect productivity by increasing or decreasing the sensory capacity of the worker, but by influencing his attitude toward the visual environment. The difficulty with these preference variables is that they are susceptible to individual differences. A preferred condition for one individual may be an undesirable one for another. For example, in colour harmony preferences for clothing, some individuals may prefer a subdued colour combination (shades of gray); whereas others may prefer extreme hue combinations like red and green.

In many cases the distinction between or contributions from these physiological and psychological factors are difficult to delineate, but since they can both be important determiners of lighting level needs, the problem must be faced. By sensory capacity we mean the physiological capacity of the human visual system to be evaluated directly by psychophysical experiments. Psychological variables (preferences) are that aspect of the environmental influence, primarily a consequence of past learning or experience, evaluated by scaling or preferential judgement techniques. A well-known example is brightness preference. We can vary illuminance levels and evaluate the observer's reaction to the environment from several standpoints: (a) pleasantness or comfort, (b) working performance and (c) psychophysics. The type of measurement that may be involved would be preference, productivity and equal visibility, respectively. As we go from (a) to (c), the relative importance of past experience or learning decreases, with the importance of physiological contributions increasing. That is, preference is significantly more affected by past experience than contrast equality measurements, which are principally measures of sensory capacities. This idea is schematically presented in Fig. 2. Some will argue that (b) is equivalent to (c). Having subjects rate targets of different contrast

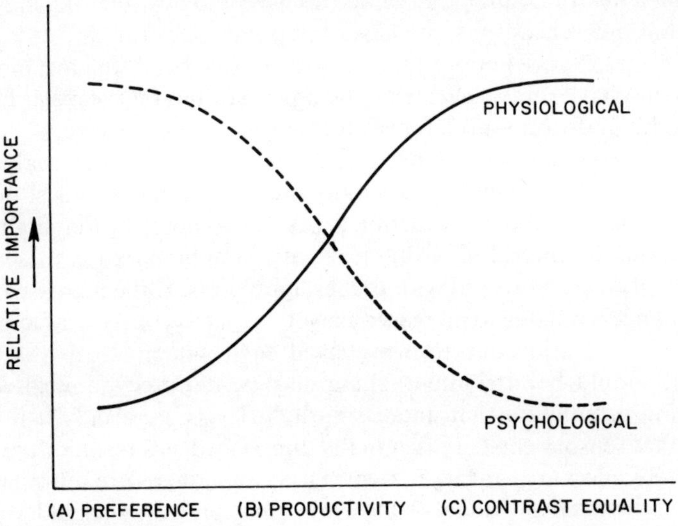

FIG. 2. Schematic demonstration of the relative importance of subjective (psychological) and objective (physiological) factors in visual performance evaluations.

as a function of luminance for legibility on a scale of 10 from good to poor, will give indications of the relative goodness of the targets. The difficulty with this approach is that we have no way of separating that portion resulting from the subject's brightness preference and that portion due to the subject's estimate of image clarity. We can be more specific and instruct the subject to use as his criterion the apparent contrast of the detail seen against its background and base his rating on this criterion. But this is a psychophysical evaluation of contrast equality using scaling techniques. This same measure can be obtained by direct comparison studies, which will lead to less variability than scaling techniques. In the latter the reference standard is a mental 'frame of reference'; in the former a direct comparison is made with a reference standard stimulus seen adjacent to the test stimulus. That is, in the second case two stimuli are directly compared for equality, whereas in the first case the given stimulus is compared against a mental reference standard. The observer is required to decide what his standard is, and keep this in his memory, which will, of course, contribute to response variability.

We have seen that performance is a function of physiological and psychological factors. Response from physiological factors can generally be predicted with confidence, while the psychological factors show large individual differences, in some cases the preferences running to opposite extremes making predictions difficult. On the one hand we have biological factors that have similar effects on most people, and on the other hand we have highly individual preferences formed by past experiences.

All this does not imply that preferences or other subjective variables need not be considered in lighting. The only precaution that we insist on is the awareness that the above distinction exists. Let us not state that higher light levels result in increased visual capacity simply because increases in productivity were observed with higher light levels. If the increase in work performance is not due to increased sensory capacity, but other factors, e.g. enhanced motivation due to a preferred environment, then it is that to which it should be attributed. There may be more economical ways of increasing production than increasing light levels, especially if it can be shown that sensory capacity is actually diminished as a result of the higher level. Preferences are to a large extent learned, and therefore subject to large variations between individuals. One worker may prefer high light levels (past experience indicated that he was able to see better when the light level was increased), whereas another may prefer lower levels (influenced by his desire to conserve energy).

What is needed is a study in which visual performance is investigated in the psychophysical sense, under laboratory controlled conditions, but at suprathreshold levels. It is also essential that subjective contributions, e.g. brightness preference, mood, motivation, etc., as well as other non-visual components (motor skills, intelligence, etc.) be controlled or at the least minimised. The laboratory experiment should be of the form: For a definitely identifiable task (non-threshold) is legibility greater under illuminance level A or B? The criterion for acceptance should meet the requirements or needs of the user.

Yonemura and Kohayakawa[10] did a study which fulfils the above requirements. They had their observers examine sinusoidal and square-wave gratings for equality of contrast at near threshold and suprathreshold levels. Adjacent grating patterns were identical in all respects except for mean luminance. The methodology differs significantly from the threshold technique, in that the grating patterns were visible at all times. The observers were asked to vary the contrast of one grating pattern seen simultaneously adjacent to another until they appeared to have equal apparent contrast. For target near threshold, just barely visible, the data resembles the classical threshold data. As luminance is increased the contrast required for equality of contrast decreases. However, a departure from classical data was observed in the 'equality of apparent contrast' contours for suprathreshold contrast levels. Instead of a monotonically decreasing function, the equal apparent contrast contour initially showed a decrease in required contrast as luminance was increased, but further increases resulted in a reversal. That is, there was an optimum luminance level, above or below which the apparent contrast of the object required for equality increased. See Figs. 3 and 4.

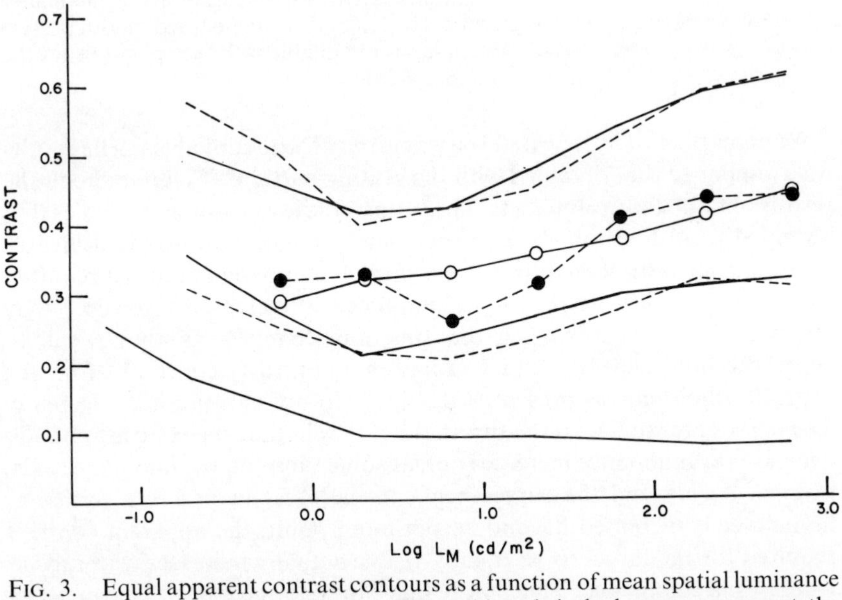

FIG. 3. Equal apparent contrast contours as a function of mean spatial luminance for a 3·9 cyc/deg. sine-wave grating. The solid and dashed curves represent the results for two different observers. The curves with the circled data points are for surround luminances maintained at the mean spatial luminance of the grating with the higher luminance.

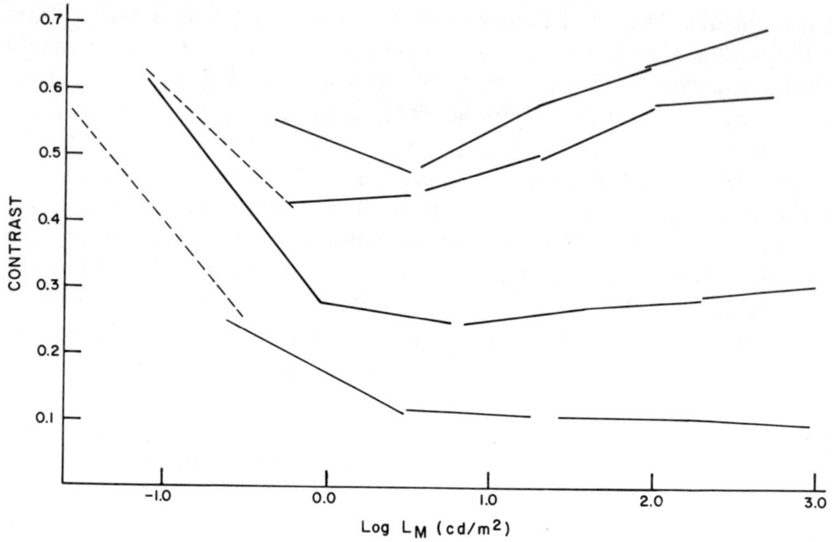

FIG. 4. Equal apparent contrast contours as a function of mean spatial luminance for a 3·9 cyc/deg square-wave grating. The left ends of the dashed lines represent contrasts which were at the maximum levels obtainable with the apparatus for the given condition.

Yonemura *et al.*[11] repeated this experiment with alphabets. The results were similar to that obtained with the grating patterns. Figure 5 shows the results obtained with alphabets. The form of the lowest curve resembles that obtained with threshold experiments. The curve monotonically decreases as luminance is increased; in other words, the apparent contrast required for equal visibility decreases as luminance is increased. However they observed that the form of the function changed as they went to suprathreshold levels. Equal apparent contrast contours for the suprathreshold curves presented in Fig. 5 do not continuously decrease. The curves are similar to the threshold curves in that the contrast initially decreases as luminance increases but, at some point on the luminance axis, reversal occurs and the curves begin to increase. In practical terms, as luminance is increased beyond an optimum point, the apparent contrast required for the target to be equally visible actually increases with further increases in luminance. Thus, more light does not necessarily mean better sight. When the reading material is of low contrast, more light leads to enhanced visibility or legibility. However for many common tasks, e.g. pen-written tasks, typewritten material, reproductions, etc., the popularly

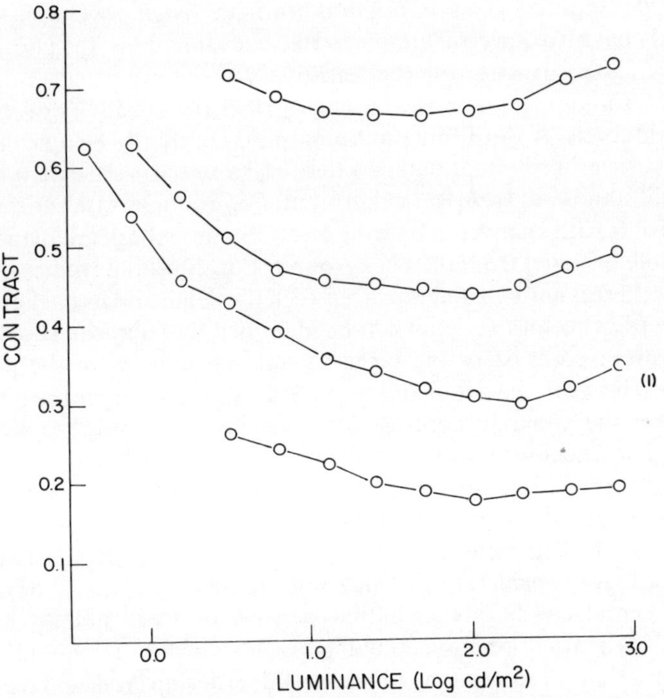

FIG. 5. Equal apparent contrast contours as a function of background luminance averaged over four subjects (curve 1 is the result for a single subject) and three letter sizes.

accepted notion that more light equals better sight, cannot be indiscriminately applied. For these tasks there is an optimum luminance level above which legibility decreases with further increases in luminance. The observation that at low contrast levels, near threshold, the equal contrast contours are of a form similar to the classical threshold monotonic function indicates that changes in the form of the function at suprathreshold levels are not due to differences in experimental procedures. Using the same apparatus, targets and procedure, the near threshold functions are monotonically decreasing in agreement with other techniques used to investigate threshold visual performance.

Does the previous discussion mean that we must discard the current basis for recommending levels of illumination? No, but it does indicate that a single visual performance criterion function is not sufficient. Two, probably three, separate functions representing different levels of task difficulty should be adopted. This is not unknown in visual psychophysics. We presently have two spectral luminous efficiency functions, the photopic and scotopic. The former is for the sensitivity of the eye when evaluated at suprathreshold levels, and the latter describes the sensitivity of the eye at threshold levels. A third function (mesopic) is currently being considered.

The basic criterion that makes a task easier to see is the extent to which the detail stands out from its background. For example, tasks are said to be more visible with changes in lighting levels, because the details can be more easily differentiated from the background. For threshold studies the test is whether there is enough difference between the detail and its background so that the presence of the detail can be identified. In suprathreshold studies the apparent contrast of the target is nothing more than the perceived difference between the detail and its background. The larger the difference, the higher the apparent contrast. A larger detail is 'easier' to see than a smaller one because it stands out more from its background. In short, apparent contrast is assumed to be the key to ease of seeing. Whatever the variable, contrast (hue, saturation or luminance), size, blur, etc., the criterion associated with the visual task is how well the detail can be seen.

The primary consideration in lighting should be 'goodness' of task and environmental conditions such that performance will not be adversely affected even after long sustained periods of work. It should also be recognised that some tasks will fall below this criterion level, and the eye will be functioning at less than desirable levels. Techniques can be used to improve the task itself, but in most cases improvements in lighting will not bring the task to an acceptable level. There will always be difficult tasks. These will be represented by a near threshold Visual Performance Criterion Function (VPCF), but we need a VPCF that prescribes a visibility level for suprathreshold tasks. From experimental data we can obtain a series of equal contrast contours, but this does not resolve the problem of which equal contrast contour should serve as the VPCF.

The criterion for a suprathreshold task can be obtained by the following procedure. At a given luminance level contrast is varied, and subjects are asked to identify that contrast level for a standard task which they will accept as the minimum acceptable level for sustained work. They are then asked to choose that level which they feel is 'definitely' acceptable. Although the absolute 'just' and 'definitely' acceptable levels may vary for the different observers, there is likely to be a region of common overlap, as observed in thermal comfort studies. This overlapping region can be chosen as the equal contrast contour to be used as the VPCF. The equating of any test task to this VPCF will follow the currently used procedure[12] with one exception.

Instead of determining the threshold contrast for the test task and evaluating it relative to the threshold contrast for the standard task, the contrast of the standard task will be varied until it matches the contrast of the test task. That is, visibility level is nothing more than the apparent contrast of the test detail as seen against its background, relative to the contrast of the standard task. The standard task and the test task are compared for equality of contrast by varying the contrast of the standard task under spherical illumination (i.e. directionless lighting, as in an integrating sphere) until it is equal to the apparent contrast of the test task also observed under spherical illumination. For example, the test task may be a page from a letter, a newspaper, book, etc. The letter sample is seen adjacent to the standard task (4 bar black-white lines), both illuminated under spherical illumination (the standard lighting system) and the contrast of the 4 bar task is varied until the perceived contrast is equal to the letter task. The background luminance for both tasks is kept constant during the comparison. As in the current threshold technique, it is assumed that the visibility of the test task is equal to the standard task regardless of the form or size of the target. The VPCF for the test task is that equal contrast contour that includes the contrast luminance value of the standard task, at the point where it matches the test task.

PHYSICAL CORRELATES

It was suggested earlier that the concept of blur has been ignored in studies on lighting and visual task performance. We further pointed out that for suprathreshold levels, where goodness of seeing (clarity or equal apparent contrast) is the dependent variable, the sharpness or definition of the detail becomes an important parameter. We further saw how the edge gradient can influence the apparent contrast of visual details. Thus, any system used to describe the visual task quantitatively should include the physical correlate of blur as well as size, contrast, luminance and time.

Modulation Transfer Function

The concept of Modulation Transfer Function (MTF), including the use of spread functions and Fourier transform techniques, may be the appropriate means for a quantified description of visual tasks, especially if the sharpness of detail is to be considered. MTF is commonly used in optics for the assessment of the performance of optical components. It is used in the field of image enhancement, that is in the sharpening of degraded images. The concept of spread function appears to be useful in modelling the role of neural inhibitory and summative effects in the visual process. Application of Fourier techniques in investigations and quantification of the visual process is discussed by Cornsweet.[13] The MTF considers the classical parameters in visual psychophysics: contrast (modulation), size (spatial frequency), and luminance as well as blur (spread function). Time can also be considered in the sense of the *D* function (time × intensity), but for most real world tasks this variable is greater than the critical duration for visual sensory processes. Critical duration is defined as that temporal period beyond which viewing time does not result in increased visibility, unless search is involved.

Edge Gradient

When a uniformly illuminated field is cut abruptly by an adjacent uniformly dark field, we have a 'knife edge'. The luminance distribution across the edge separating these two fields is the edge gradient. The physical correlate of blur must be some property of the trace (by a microdensitometer) of the edge of the image, that is of the edge gradient. Figure 6 is a schematic presentation of an edge. The ordinate is given in terms of optical density rather than transmittance or reflectance since the response of the eye to luminance is essentially logarithmic. The edge gradient distance X_a to X_b is the distance perpendicular to the contour. Figure 7 presents the luminance distribution of three real edges, a master photograph and two degraded levels of the master. The edge gradients were obtained by scanning the edge

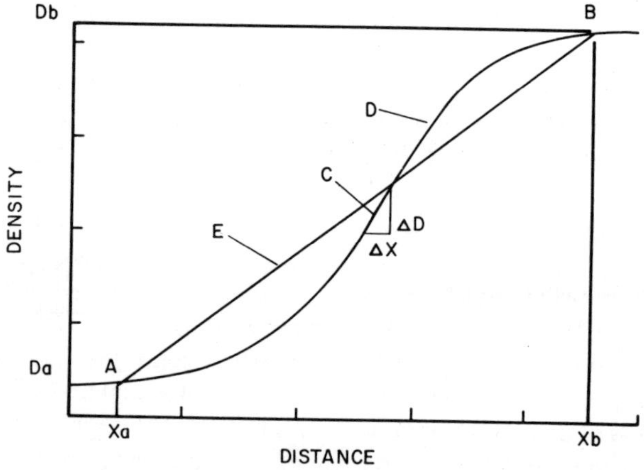

FIG. 6. Hypothetical microdensitometer traces of two edge gradients E and D.

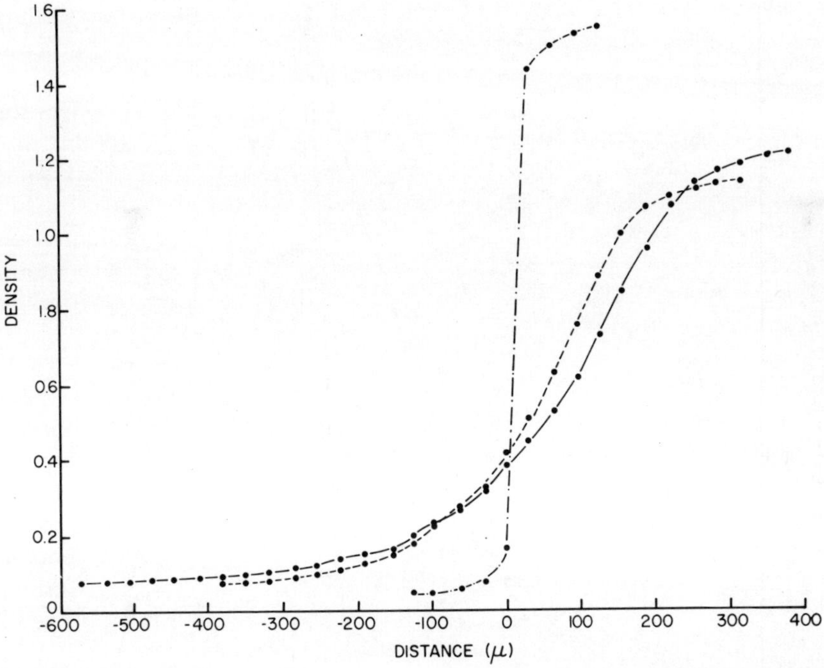

FIG. 7. Microphotometry of knife edges. The dash-dot curve is the original knife edge, the dashed curve is for degradation level 9 and the solid curve is that obtained under degradation level 11. The point of maximum slope does not occur at zero because the plotted values are log transforms (density) of the luminance data.

with a microphotometer. The luminance readings were continuously recorded on an x–y recorder. The master knife edge (dash-dot) has a sharp edge, but a significant change in the shape of the edge gradient is observed for two degradation levels, labelled 9 (dash) and 11 (solid), representing two degrees of blur. The portion of the edge gradient with the largest slope occurs at $x = 0$.

Line Spread Function

The edge gradient is directly related to the MTF by its first derivative, the Line Spread Function (LSF). Since the first derivative is the slope of the edge gradient at x, a plot of the slope against x is the LSF:

$$\text{LSF} = A(x) = \mathrm{d}D(x)/\mathrm{d}x \qquad (1)$$

Although the fundamental parameter of modulation transfer function is

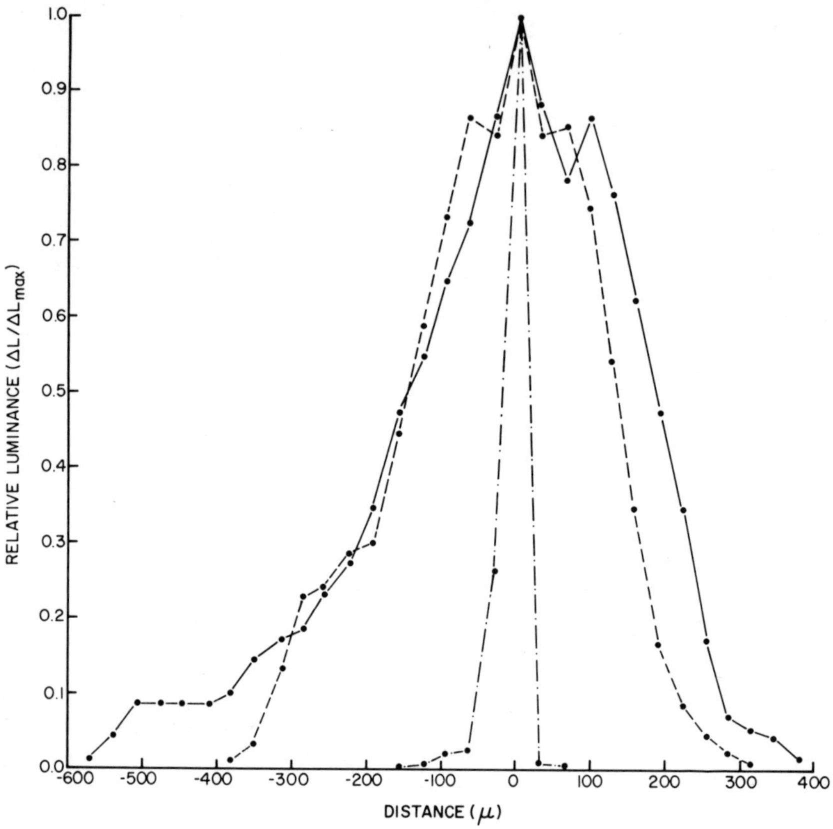

FIG. 8. Line spread functions for the original knife edge (dash-dot), degradation level 9 (dash) and degradation level 11 (solid).

the spread of a point source rather than a line spread, in practical microphotometry a slit aperture is used to increase the energy available for measurement. Therefore, the integral of the point spread function in one direction, that is the LSF, is utilised. The LSFs obtained from the edge gradients shown in Fig. 7 are presented in Fig. 8.

Convolution

The luminance distribution of the image of any object whose luminance varies in one direction can be determined by convolution of the LSF with the object luminance distribution. Let $G(x)$ be the luminance distribution as a function of distance x for a one-dimensional object. Take a relatively simple distribution, the cosine distribution:

$$G(x) = b_0 + b_1 \cos 2\pi v x \quad (2)$$

where b_0 is the mean luminance, b_1 the amplitude and v the spatial frequency, as shown in Fig. 9. Given the luminance distribution of the object, $G(x)$, and the LSF, $A(e)$, the luminance of the image at point x is obtained by:

$$L(x) = \int_{-\infty}^{\infty} A(e) G(x - e)\, \mathrm{d}e \quad (3)$$

where e is the distance from the point in the test object (x) at which the luminance is to be determined. This operation is called the convolution of the object function by the line spread function.

For the cosine object distribution we have:

$$L(x) = b_1 \int_{-\infty}^{\infty} A(e)\, \mathrm{d}e + b_1 \int_{-\infty}^{\infty} A(e) \cos 2\pi v(x - e)\, \mathrm{d}e \quad (4)$$

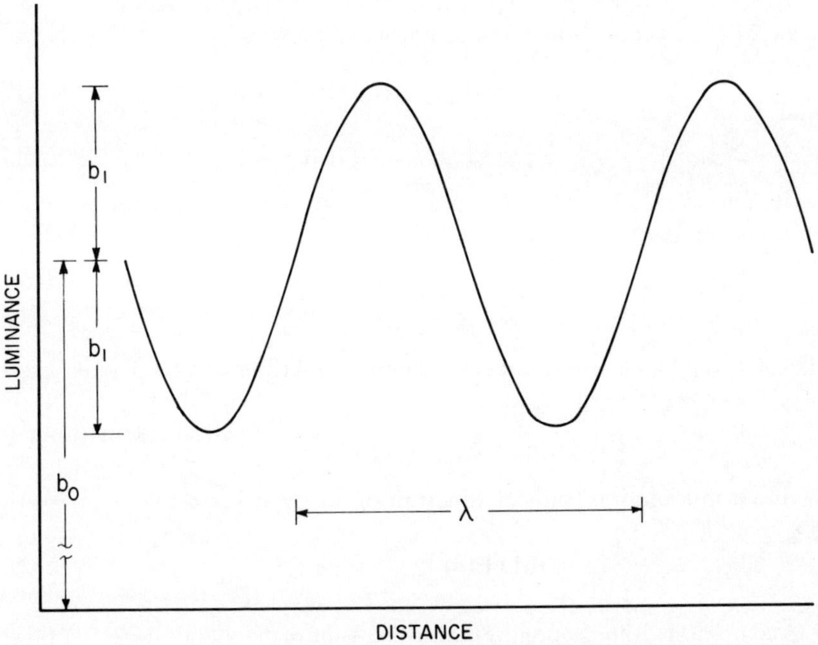

FIG. 9. Sine-wave distribution illustrating definition of modulation $= b_1/b_0$.

We normalise eqn. (4) by dividing through by $\int_{-\infty}^{\infty} A(e)\,de$, the area under the LSF. The normalised equation may be rewritten as:†

$$L(x) = b_0 + b_1|A(v)|\cos(2\pi v x - \phi) \tag{5}$$

where:

$$|A(v)| = [A_c^2(v) + A_s^2(v)]^{1/2} \tag{6}$$

is the Fourier transform of the LSF, and

$$A_c(v) = \frac{\int_{-\infty}^{\infty} A(e)\cos 2\pi v e\,de}{\int_{-\infty}^{\infty} A(e)\,de} \tag{7}$$

and

$$A_s(v) = \frac{\int_{-\infty}^{\infty} A(e)\sin 2\pi v e\,de}{\int_{-\infty}^{\infty} A(e)\,de} \tag{8}$$

The subscripts c and s refer to cosine and sine, respectively. ϕ is the phase difference between object and image. The phase angle vanishes for a symmetrical line spread function,

$$|A(v)| = A_c(v). \tag{9}$$

In eqn. (5), for a symmetrical LSF, $L(x)$ is a maximum when $\cos(2\pi v x) = 1$, and is at a minimum when $\cos(2\pi v x) = -1$. Therefore:

$$L(x)_{\max} = b_0 + b_1|A(v)| \tag{10}$$

and

$$L(x)_{\min} = b_0 - b_1|A(v)| \tag{11}$$

In Fig. 9, $L_{\max} = b_0 + b_1$ and $L_{\min} = b_0 - b_1$. Using these, object modulation (contrast) is defined as:

$$M = \frac{L_{\max} - L_{\min}}{L_{\max} + L_{\min}} = \frac{b_1}{b_0} \tag{12}$$

Then, using eqns. (10) and (11), image modulation can be defined as:

$$M' = \frac{[b_0 + b_1|A(v)|] - [b_0 - b_1|A(v)|]}{b_0 + b_1|A(v)| + b_0 - b_1|A(v)|} = \frac{b_1}{b_0}|A(v)| = M|A(v)| \tag{13}$$

giving a modulation transfer function of

$$\mathrm{MTF}(v) \equiv \frac{M'}{M} = |A(v)| \tag{14}$$

† A complete derivation of eqn. (5) and a discussion of the phase transfer function is given by Lambert.[14]

The LSF, $A(e)$, cannot be expressed analytically in many cases. Therefore this function must be considered point by point and the Fourier transform integration must be treated as a summation process. Using eqns. (7) and (9)

$$\mathrm{MTF}(v) = \frac{\sum_{i=n}^{m} A(e_i)\cos 2\pi v e_i \Delta e_i}{\sum_{i=n}^{m} A(e_i)\Delta e_i} \tag{15}$$

where i ranges from the smallest measurable n on one side of the spread function to the smallest measurable value on the other side. The MTFs derived from these spread functions using eqn. (15) are presented in Fig. 10. In Fig. 10 the abscissa is given as frequency in cyc/mm and cyc/deg., since in most publications on human response to sinusoidal gratings the data are generally presented as cyc/deg or in cyc/mm on the retina. The MTFs were not corrected for the slight gradient in the master knife edge (see Fig. 7).

Figure 10 tells us that when a cosine pattern is degraded to the extent indicated by the edge gradients shown in Fig. 7, the loss in contrast for patterns larger than 0·25 cyc/mm is small. The solid curve represents the contrast loss to be expected from a cosine pattern when a system degrades the image to the extent given by the edge gradient indicated by degradation level 11. For a pattern with relatively large dark–light bars or low spatial frequency the loss in contrast is small, but for frequencies exceeding 3 cyc/mm the pattern is seen as an homogeneous field. For a pattern with a frequency of 1·0 cyc/mm the contrast will be approximately $\frac{3}{5}$ of the original contrast (see Fig. 10). As might be expected the loss in contrast is not as great for edge gradient 9 as it is for 11.

We saw that the variables considered in task description in lighting have a physical counterpart in MTF that is amenable to quantitative treatment. Granger and Cupery[15] suggest that a single number index may be obtained by integrating the MTF over a limited frequency band. They found that this MTF-based index of image quality correlated highly (0·988) with film print quality rank. The quality of the prints was varied by simple defocus, contrast loss, Gaussian degradation, simulated central obstruction of lens aperture, film adjacency effect and astigmatism. There appears to be a basis for considering the MTF as a physical correlate for task description in lighting. More research and validation experiments are needed, especially with tasks encountered in the real world.

The subject of Equivalent Sphere Illumination (ESI) and Contrast Rendition Factor (CRF) as a measure of lighting system effectiveness is

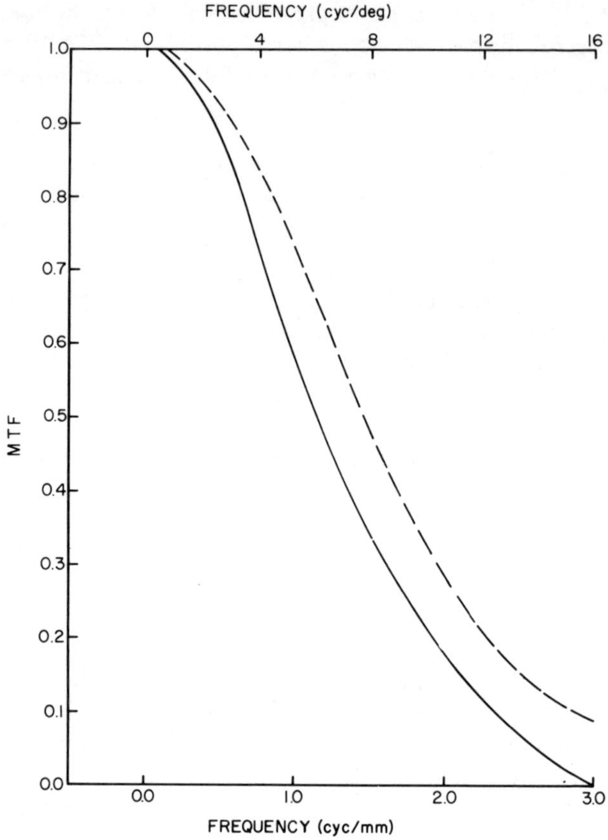

FIG. 10. The Modulation Transfer Function for a cosine pattern convoluted with the Line Spread Functions of Fig. 8. The solid curve is for degradation level 11 and the dashed curve for level 9. A viewing distance of 12 in (30 cm) was used in the frequency axis given in cyc/deg.

discussed in Chapter 1. Basically what is involved in ESI is the contrast rendering capability of the lighting system relative to the contrast obtained from a spherical lighting system, the real-world analogy being a large totally diffuse lighted ceiling. CRF is the ratio of two contrasts, the contrast of the task under a test lighting system (spherical) relative to the contrast of the same task under a reference lighting system. MTF, as we saw earlier, is also the ratio of two contrasts, the modulation (contrast) of the output image relative to the input image modulation. They are both measures of changes in luminance contrast. In lighting, CRF primarily evaluates the change in contrast resulting from veiling reflection (regular reflection superimposed on diffuse reflection that causes decrease in contrast). In the MTF context, this is the same as superimposing a grating of zero frequency, $b_1 = 0$ in eqns. 10 and 11, on an existing grating pattern. The advantage of adopting terms firmly established in optics is that the transfer of findings from a highly investigated area to lighting will be more readily appreciated and utilised.

REFERENCES

1. Fox, W. R., in *Form Discrimination as Related to Military Problems*, (ed. J. W. Wulfeck and J. H. Taylor) 1957, Publication 561, National Academy of Science–National Research Council, Washington DC.
2. Perrin, F. H., *J. Soc. Motion Picture Television Engrs.*, 1960, **69,** 151.
3. Brainard, P. W. and Caum, K. B., 'Evaluation of an Image Quality Enhancement Technique', AMRL-TR-65-143, Sept. 1965, Aerospace Medical Research Laboratories, Wright–Patterson Air Force Base, Ohio.
4. Ratliff, F., in *Sensory Communication*, (ed. W. A. Rosenblith) 1961, MIT Press, Cambridge, Mass.
5. Ratliff, F., *Mach Bands: Annotative Studies on Neural Networks in the Retina*, 1965, Holden-Day, Inc., San Francisco.
6. Ratliff, F., *Proc. Am. Phil. Soc.*, 1971, **115,** 150.
7. O'Brien, V., *J. Opt. Soc. Amer.*, 1958, **48,** 112.
8. Luckiesh, M., *Illum. Engng.*, 1948, **43,** 395.
9. Tinker, M. A., *Illum. Engng.*, 1948, **43,** 866.
10. Yonemura, G. T. and Kohayakawa, Y. 'A New Look at the Research Basis for Lighting Level Recommendations', NBS Building Science Series 82, March 1976, US Government Printing Office, Washington, DC.
11. Yonemura, G. T., Benson, W. M. and Tibbott, R. L., 'Levels of Illumination and Legibility', NBS Interim Report. In press.
12. Kaufman, J. E. (ed.), *IES Lighting Handbook*, 5th ed., 1972, Illum. Engng Soc., New York.
13. Cornsweet, T. N., *Visual Perception*, 1970, Academic Press, New York.
14. Lambert, T. L., *J. Opt. Soc. Amer.*, 1958, **48,** 490.
15. Granger, E. M. and Cupery, K. N., *Phot. Sci. and Eng.*, 1972, **15,** 221.

Chapter 3

UNDER THE INFLUENCE OF LIGHTING

P. R. BOYCE
B.Sc., Ph.D., M.I.S., M.C.I.B.S.
Research Officer, Electricity Council Research Centre, Chester, UK

SUMMARY

Lighting can be used to manipulate the impression an interior gives. It is the factors controlling this capability and the methods of identifying them that are of interest here. The methods that have been used recently—prediction from an understanding of how visual perception works, multiple scaling techniques followed by multivariate analysis, and simple observations of behaviour—are summarised and evaluated. The knowledge of the factors controlling the impression lighting gives to an interior, which has been obtained by these methods, is discussed and some future possibilities are considered.

PAST, PRESENT AND FUTURE

If the lighting of a space is designed to the recommended standards of illuminance, uniformity and glare control, and uses one of the suggested light sources[1] it can hardly fail to be comfortable to the occupants. This is not to say it will be suitable and pleasant. It may well be bland, boring and bright but it will not be uncomfortable. This is not a situation to be ridiculed but neither is it a justification for complacency. It is no mean achievement to banish discomfort from lighting but how much better if lighting could make a more positive contribution. And it can. An interior may be perceived as dramatic, depressing, exciting, relaxing, interesting, workmanlike, etc., depending on the lighting. It is the use of lighting as a tool to manipulate the impression an interior gives that allows us to talk of the influence of lighting. Over the last few years a number of new approaches have been used to identify how lighting conditions affect people's responses to interiors. These approaches form the subject material of this chapter. The aim is to summarise and evaluate them, in the hope that lighting engineers and researchers will be encouraged to consider lighting systematically as a positive influence in the design of interiors.

Before describing the new approaches, it is worthwhile examining the methods of assessing lighting that have brought us to our present state. Almost all the quantitative criteria used in lighting design, and which serve to avoid discomfort, have been derived from simple correlation studies. In essence, this approach consists of relating some subjective judgement, usually expressed on a rating scale, to a chosen physical descriptor of the lighting. The physical descriptor is often selected not for its importance to the person making the assessment but because it is measurable and/or can be used in the design process, e.g. illuminance on a horizontal plane. The subjective judgement is usually chosen either to be very general, e.g. a satisfactory/unsatisfactory rating scale, or clearly linked to a physical measure, e.g. a bright/dim rating scale. If the physical quantity correlates well with the subjective judgement it is taken to be a physical index of the psychological variable represented by the judgement and can then be used as a criterion for the design of the physical environment. Various forms of this correlation approach have been adopted in producing criteria for illuminance, uniformity and discomfort glare[2,3] and in a study of perceived spaciousness.[4]

Having designed lighting installations with these criteria the reality has sometimes been examined by field appraisals.[4,5] These usually involve taking a group of people into an installation and asking them to make judgements in a way similar to that used in the laboratory. Therefore the whole derivation and test process forms a closed system. The lighting criteria are developed from studies in laboratories using a limited range of physical variables and subjective judgements and the recommendations are tested in the field using the same type of assessment. Certainly there is little opportunity for people to assess the lighting in their own terms, rather than those imposed by the experimenters. This whole approach has been criticised by Hawkes.[6] His objections to the method are twofold, first it assumes that because a high correlation can be obtained between a physical measure and a subjective judgement the psychological variable represented by the judgement is significant in people's everyday experience, and second, it assumes that the rating scales on which the judgements are usually made represent independent psychological variables or dimensions. To these criticisms may be added the generalisation that the whole approach starts from the physical end of the problem rather than the people end. The first step is usually to introduce a new physical measure of the luminous environment, rather than identifying how people perceive interiors.

These comments suggest that whilst the correlation approach may be suitable for validating some proposed lighting measure, it can produce only a limited understanding of the ways in which lighting influences people's perceptions of an interior. Therefore it seems unlikely that further research of this type will prove capable of promoting lighting practice from its present negative state of avoiding discomfort to a more positive role in the design of interiors. To do this a more fundamental approach is needed, one which starts by considering people rather than photometers.

AN INSIGHT FROM VISUAL PERCEPTION

A basic understanding of the lighting requirements necessary to manipulate the impression an interior gives can be reached from an appreciation of how visual perception operates.[7] Visual perception is dependent on the complete eye/brain system. When the eye focuses on a scene the retinal organisation tends to emphasise the effect of contrast and to enhance sharp gradients of luminance. From such information and the data-bank of previous objects viewed, mental constructions of what the scene is are produced and acted upon. By adulthood the brain is furnished with a store of expectations about the visual environment with which incoming data are compared. What we see is usually the best fit between these expectations and the incoming data. It can be conveniently classified into one of three modes of appearance. These are: (a) the surface mode in which the stimulus is perceived as a coloured surface of an object, (b) the illuminant mode in which the object is perceived as self-luminous and (c) the aperture or film mode which is typified by a uniform expanse of blue sky. The modes of appearance do not describe the object itself but only how it is seen, the same object being capable of appearing in different modes depending on the lighting conditions. Each mode is associated with the presence or absence of specific perceptual cues. The surface mode of appearance is experienced when we can see the texture, sheen, etc. typical of a surface, the sharp boundaries normally associated with objects and the relative luminances and colours of the object and background. However if the luminance of the object is much greater than the background it will move into the illuminant mode. The aperture mode of appearance is experienced when spatial cues are absent, as occurs in conditions of very low illuminance.

The existing criteria for good lighting of interiors include illuminance on a specified plane with a minimum spatial uniformity, the vector/scalar ratio, illuminance ratios for different surface combinations, and the use of light sources with reasonable colour properties. If the illuminances in an interior are kept within the limits prescribed the illuminant mode is avoided. If sufficient illuminance and the right vector/scalar ratio are provided, the microstructure of surfaces is revealed so the surface mode is maintained. Lamps with good colour properties ensure good chroma for colours and good contrast between object and background and hence support the surface mode. There can be little doubt that the present lighting criteria, consciously or unconsciously, tend to ensure that objects are seen in the surface mode. Further it would appear that by emphasising contrasts and enhancing sharp luminance gradients the visual system operates in a way which is well suited to preserving the surface mode.

An alternative way of considering how visual perception operates is in terms of brightness and colour constancy. By constancy is meant that the brightness and colour of objects stay more constant than they physically should. A black object near a window will not look brighter than a white object some distance from the window even though the luminance of the former is greater than the latter. Similarly a coloured wall lit by daylight at one end and an electric light at the other will not appear to be two different colours. It has been observed that existing lighting criteria also serve to maintain constancy as well as the surface mode of appearance.[8] Now constancy, if it means anything, means that objects look the same under a wide range of lighting conditions. Conversely, objects will look different when constancy breaks down. This implies that an understanding of the lighting conditions which control constancy, and the mode of appearance, can give some indication of how the perception of an interior can be manipulated. Both Jay[7] and Lynes[8] discuss this topic in more detail, but the possibilities can be illustrated by considering display lighting. Such lighting should be exciting and dramatic. This can often best be achieved when constancy starts to break down and the surface mode of appearance changes to the illuminant mode. This state can be reached when the relevant objects are brightly lit and those which are not of importance merge into the background. Similar principles apply to flood-lighting. It is often most effective when the lit objects lose their surface quality and begin to appear self-luminous. Thus lighting for effect depends on large contrasts in the field of view, produced by brightly lighting objects which are of interest in themselves. Such lighting can be interesting even at a very low overall illuminance provided the gradients of luminance are sharp. At the other extreme if the gradients of luminance are small then the visual system can adapt as it scans the visual field. This will usually result in an impression of gloom in the interior, which can readily be relieved by a few sources of high luminance scattered around.[7]

It should now be apparent that an understanding of visual perception can delineate the boundary conditions for different types of lighting, depending on the function they are to serve. But this is not enough on its own. To pursue an analogy, visual perception will help us to mark out the football field but more detail is necessary if we are to understand the game.

WHAT DO YOU MEAN—GLARING?

One way to provide this greater detail is shown by the work of Kimmel and Blasdel.[9] They asked 112 subjects to rate the luminous environment in 10 different library reading rooms on a number of 7-point rating scales. The actual rating scales were concerned with features of the lighting that had been used in the design, e.g. glare from lights, quantity of light, colour of light. In this respect the method differs little from the correlation approach described earlier. However it does differ when it comes to the method of analysis.

The ratings given were not averaged and plotted against some apparently relevant physical measure. Rather the ratings were used to form a matrix of differences of scores between the various libraries on the same scale for the same subjects. From the difference score matrices formed in this way the minimum number of independent dimensions necessary to explain the pattern of differences was identified by a form of the Multi-Dimensional Scaling (MDS) analysis.[6] Thus this method does not involve physical measures but uses only the interrelationship of the subjects' judgements to discover the basis on which lighting is being assessed. Figure 1 is an example of the sort of information that is available from this approach. It shows the position of the various reading rooms on the plane formed by two identified dimensions of assessment, i.e. illumination and glare from lights. There is a consistent trend for poorer glare assessment to be associated with higher perceived illumination, which is interesting enough, but the three installations which do not follow the trend are even more informative. Installation B is a high luminous ceiling which has less glare than expected.

Installation H is a high luminance downlighting installation which has more glare than expected. Installation MT should have had a much lower glare level according to the luminance of the luminaires but actually it was the worst. The authors suggest this was because the luminaires were arranged in a very striking pattern on the ceiling. All three rogue results can be explained if discomfort glare involves an element of distraction towards striking and irrelevant high luminance stimuli, the greater the distraction the more the discomfort glare. This thought has been raised from other sources[10] and it is interesting to have it suggested from field data.

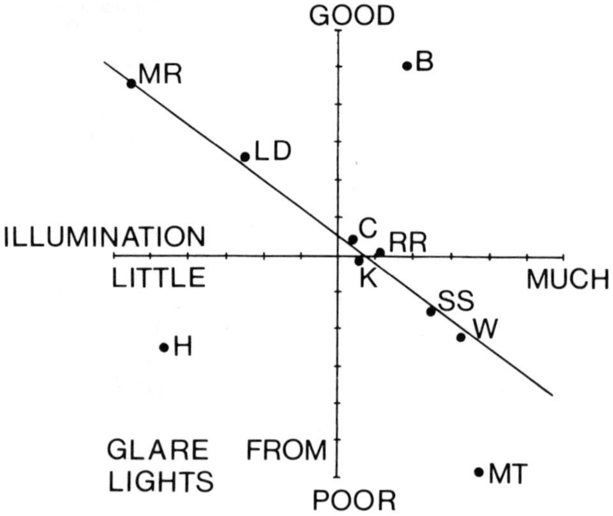

FIG. 1. The positions of the 10 library reading rooms on the plane formed by the dimensions of illumination and glare from lights.[9] (Reprinted from the January 1973 issue of *Journal of the Illuminating Engineering Society* with permission of the Illuminating Engineering Society of North America.)

It is obvious that there is much to be learnt from this approach[9] since it offers the possibility of identifying the physical factors which are important to people's perception of an interior from assessments of complete real interiors. However it still does not overcome all the criticisms of the correlation approach. Although the MDS analysis ensures that all the dimensions used are independent, the questions still refer to the prevailing criteria of lighting. There is little opportunity for the observer to speak for himself.

THE OBSERVER AS KING

One research team that has tried to do this is Flynn *et al.*[11] These researchers demonstrated a conference room lit in six different ways to 96 observers. Each installation was rated by each observer on 34 different rating scales. The rating scales were not related to lighting but were concerned with the observer's impressions of the room. The data collected was then subjected to a factor analysis.[12,13] This is a statistical technique designed to bring order out of chaos. The method can take a great mass of such rating scores and by examining the correlations between all the various scales taken in pairs can produce an estimate of the number of underlying independent dimensions upon which the observer is basing his assessment of the various stimuli, which in this case are lighting installations. Further the factor analysis will give a measure of the extent to which each individual rating scale is related to the independent dimensions. The best way to illustrate this process is by example. Figure 2 shows the six installations that Flynn *et al.*[11] used together with some physical data. Figure 3 shows the five independent dimensions on which the assessments of the room seen under the six lighting installations were based, and the individual rating scales which are most strongly related to each dimension. It should be noted that the names of each dimension (or factors as they are sometimes called) are not given by the factor analysis. They are inspired guesses produced by inspection of the scales most strongly related to the dimensions. The five dimensions found have been named 1. evaluative (i.e. pleasant/unpleasant), 2. perceptual clarity, 3. spatial complexity, 4. spaciousness and 5. formality.

Also shown in Fig. 3 are the mean ratings for each installation on each scale. From an examination of these it is possible to appreciate the richness of the information available about the observer's assessment of the room under the different lighting conditions. For a start consider the evaluative dimension, using the pleasant/unpleasant scale as it is the one most strongly related to that dimension. The two installations that are regarded as most pleasant have a common feature. They both provide a combination of lighting on the conference table and lighting on the walls. The two most unpleasant lighting installations also have something in common—they both provide diffuse lighting only on the conference table, although one had an illuminance of 110 lx and the other 1100 lx. So much for the idea that pleasantness can be a function of illuminance alone. Although the above discussion has been concerned with only one scale, an inspection of Fig. 3 will reveal a very consistent pattern for the different installations on all the scales related to the evaluative dimension.

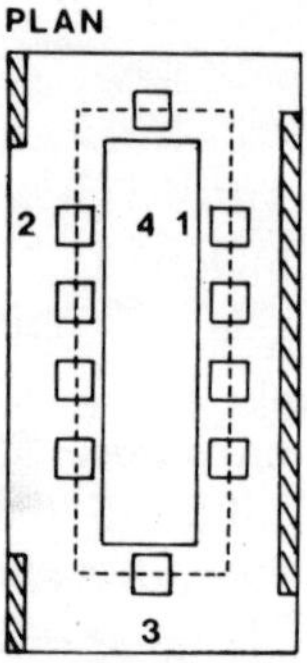

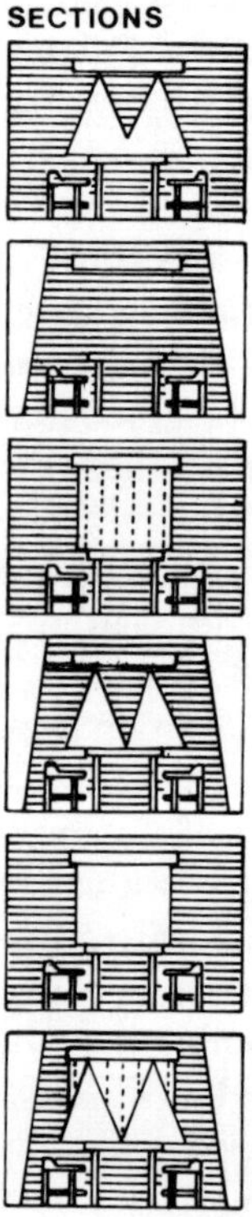

Sections	Luminance (cd/m²) see plan for position: 1 Table	2 Wall	3 Wall	4 Ceiling	Illuminance (LUX): Table
1 OVERHEAD DOWNLIGHTING LOW INTENSITY	1·2	0·3	0·0	1·7	110
2 PERIPHERAL WALL-LIGHTING, ALL WALLS	7·9	19·9	32·5	8·9	110
3 OVERHEAD DIFFUSE, LOW INTENSITY	2·9	3·7	2·6	80·5	110
4 COMBINATION: OVERHEAD DOWNLIGHTING (1) + END WALLS	1·4	1·0	26·3	1·7	110
5 OVERHEAD DIFFUSE, HIGH INTENSITY	21·9	119·9	26·7	1404·7	1100
6 COMBINATION: OVERHEAD DOWNLIGHTING (1) + PERIPHERAL (2) + OVERHEAD DIFFUSE (3)	11·6	33·5	38·3	82·2	320

FIG. 2. A room plan, and six schematic illustrations of the six lighting installations used in the conference room.[11] Some physical measurements of the lighting conditions for each installation are given. (Reprinted from the October 1973 issue of *Journal of the Illuminating Engineering Society* with permission of the Illuminating Engineering Society of North America.)

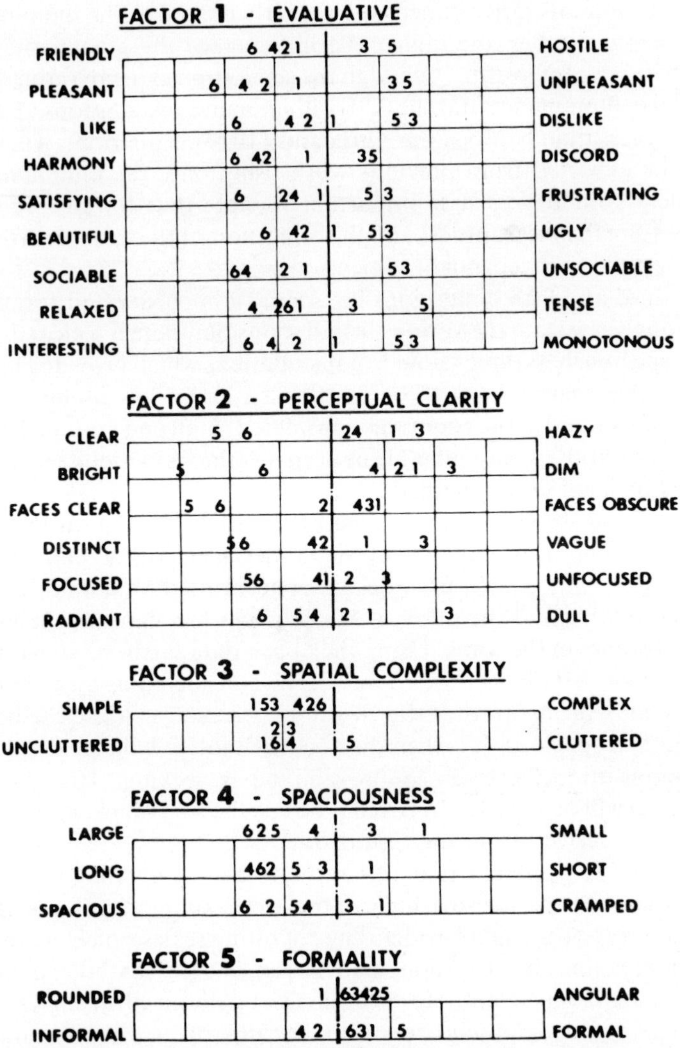

FIG. 3. A diagram of the five dimensions (factors) identified from the ratings of the conference room lit by the six lighting installations shown in Fig. 2. The rating scales most strongly related to each dimension are shown under each factor title. The mean rating for each installation on each scale is given by the appropriate installation number, i.e. installation 6 has the mean rating closest to the pleasant end of the pleasant/unpleasant scale. (Reprinted from the October 1973 issue of *Journal of the Illuminating Engineering Society*, with permission of the Illuminating Engineering Society of North America.)

The perceptual clarity dimension is clearly related to the illuminance on the conference table, the higher the illuminance the greater the clarity, brightness, etc. Interestingly enough the separation in mean rating between the 1100 lx illuminance and the 320 lx illuminance installations (5 and 6) is generally less than between the 320 lx and 110 lx installations. This is what would be expected from previous work using only the illuminance as a variable.[2] Nonetheless it is important to note that for this conference room setting the pleasantness and the brightness of the room are treated by the observers as independent dimensions.

The third and fifth dimensions i.e. spatial complexity and formality do not strongly separate the various installations but there is a clear difference for the spaciousness dimension. For installations which provided light only on the table, particularly those which also had a low illuminance, e.g. installations 1 and 3, the room was considered small and cramped while for those which lit the walls and table or even just the walls alone the room was assessed as large and spacious.

It is now possible to give an idea of the sort of question that can be answered using this data. Suppose a designer has a choice of two approaches; one which provides a concentrated illuminance on the conference table and the other which provides less light on the table but does put some on the walls. Using the above data it can be suggested that with the first installation the room will be considered as clear but rather unpleasant, whilst with the latter it will be assessed as less clear but more pleasant and spacious. It is then the designer's job to balance these various perceptions to achieve the impression he is seeking. His chances of succeeding will be greatly improved if he has the sort of installation profile that can be derived from the results discussed.

It should be apparent that this rating scale/factor analysis approach avoids some of the criticisms made of the earlier simpler methods. It allows the observers to use a much wider range of rating scales which are related to their impressions of the room and it can ensure that the dimensions obtained are independent. However it is not without blemish. The results obtained will very definitely depend on the range of conditions examined. If a variable is not presented to the observers then no dimension related to that variable will be found. For example, there is no dimension of colour in the above example and yet in general, the colour properties of the surfaces seem likely to be important variables in determining the character of an interior. Another criticism is that the rating scales are chosen by the experimenter not by the subjects. If the experimenter chooses numerous rating scales that are aimed at one dimension then this dimension will be obtained. By now there is some evidence on suitable rating scales for assessing the environment,[14] so the choice of scales is not too great a problem in practice although it is still one in principle. After all it is a form of putting words into the subjects' mouths.

Fortunately there is an alternative method which avoids some of these problems though it does not produce results with such obvious face validity as the factor analysis approach. The technique is called multi-dimensional scaling, MDS for short.[6, 15] It involves simply asking subjects to assess a range of lighting installations by judging the degree of similarity or difference between pairs of installations (cf Kimmel and Blasdel).[9] Thus no rating scales are involved in this method and the observers use their own criteria for making the judgement. Flynn *et al.*[11] used this approach to assess the six installations shown in Fig. 2 with a different set of 46 observers. The MDS procedure estimates the number of independent dimensions that are needed to explain the obtained difference ratings. These n independent dimensions form a n-dimensional space in which the individual installations are located. Flynn *et al.*[11] found that three independent dimensions formed the most parsimonious but complete arrangement to explain their data. Figure 4 shows these three dimensions with the location of the individual installations in the three-dimensional space. The three dimensions have been called peripheral/overhead, uniform/non-uniform, bright/dim. This naming is a 'bootstraps' exercise in that the three dimensions are named from a consideration of the relative positions of the installations in the space; that is relative to each other and to each dimension. Since these dimensions are identified from the installation's characteristics there is no possibility of directly evaluating the meaning of the dimensions to the subject. The most that can be done directly is to tell which installations are considered more or less different from each other, and the independent dimensions on which they differ.

However a comparison of the results obtained by this MDS approach and the factor analysis method for the six installations in Fig. 2 shows how these two methods can be mutually supporting. The three dimensions that most clearly separated the various installations in the factor analysis were perceptual clarity, evaluative and spaciousness. After choosing a rating scale representative of each of these three dimensions it is possible to do stepwise multiple regressions between the mean ratings for each installation on each scale and the three MDS dimensions. Perceptual clarity gave a correlation of 0·99 on the bright/dim MDS dimension indicating almost perfect agreement between one dimension obtained by factor analysis and one by the MDS procedure. Similarly the evaluative

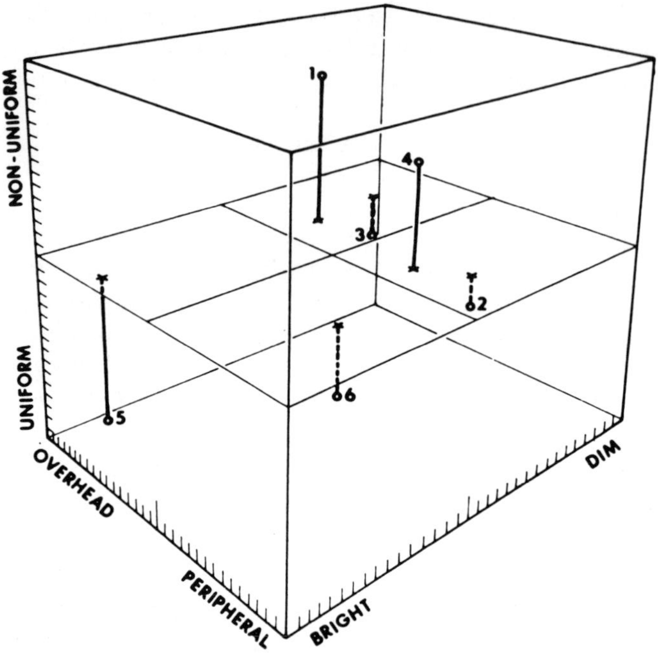

FIG. 4. The positions of the 6 lighting installations in the conference room shown in Fig. 2 in the three-dimensional space derived from the multi-dimensional scaling analysis.[11] (Reprinted from the October 1973 issue of *Journal of the Illuminating Engineering Society* with permission of the Illuminating Engineering Society of North America.)

dimension gave a correlation of 0·83 on the overhead/peripheral MDS dimension alone, and the spaciousness dimension gave a correlation of 0·69 with the uniform/non-uniform MDS dimension alone. There is therefore considerable, but not perfect, agreement between the dimensions obtained by the two methods. This agreement enables the MDS dimensions, which the subjects use to identify the differences between the installations, to be related to the subjects' impressions of the interior. For example it is clear that the extent to which the room is considered pleasant is primarily related to the existence of peripheral lighting, although since the introduction of the uniform/non-uniform MDS dimension increases the correlation with rated pleasantness to 0·92, having a non-uniform distribution of light also increases rated pleasantness. This conclusion could have been deduced from an inspection of Figs. 2 and 3 but it is nice to have it confirmed by results that involve a minimum of direction of the observer.

As both techniques lead to a similar result, which is the better? The MDS system has the attraction of leaving the observers to make their judgements of difference entirely on their own criteria but it is not as immediately understandable as the factor analysis approach, neither does it provide a set of rating scales which can be examined individually and related to the independent dimensions identified. Thus whilst the MDS technique is probably more pure it is also more difficult. The best approach appears to be to use a combination of the two methods since they are complementary.

However before advocating the methods too strongly it is worthwhile considering if the dimensions discovered and the positioning of specific installation types on these dimensions, are sufficiently stable to make the results obtained suitable for practical application. For a start, what happens in different rooms? Flynn *et al.*[16] have looked at this question in two stages, first considering rooms that have the same context for the user, and secondly, rooms with obviously different functions. Ratings were made of three different conference rooms by independent groups of subjects, one room being that used previously (room C in Fig. 5), the other two being a similar size room, but irregular in shape, and a large rectangular room. Each room was lit in five different ways, using fluorescent and incandescent sources and overhead and peripheral lighting. Factor analysis revealed four main factors which were named perceptual clarity, spaciousness, evaluative and spatial complexity. Flynn *et al.*[16] present the results in the form of mean ratings for each installation type in each room for six rating scales (Fig. 5), four being clearly representative of the perceptual clarity, spaciousness, evaluative and spatial complexity dimensions. The other two, private/public and relaxed/tense were chosen for their interest, although the latter was in fact related to the evaluative dimension. Both these additional scales show strong differences between the installations. Inspection of Fig. 5 reveals there is considerable agreement in the way the five different lighting installations in the three rooms are rated on each scale. These results imply that the type of furnishing is not as important as might be expected, subject to the limitation that the furnishings are typical of the context. The basic factor structure is very similar for room C alone and when other conference rooms are examined. Further, a comparison of Figs. 3 and 5 will show that similar lighting conditions are similarly positioned on each scale in the two experiments involving conference rooms.[11,16] It can be concluded that there is certainly some stability in the dimensions on which people assess the lighting of conference rooms and the way in which particular lighting conditions are located on those dimensions.

But what about rooms which do not have this context for the observers?

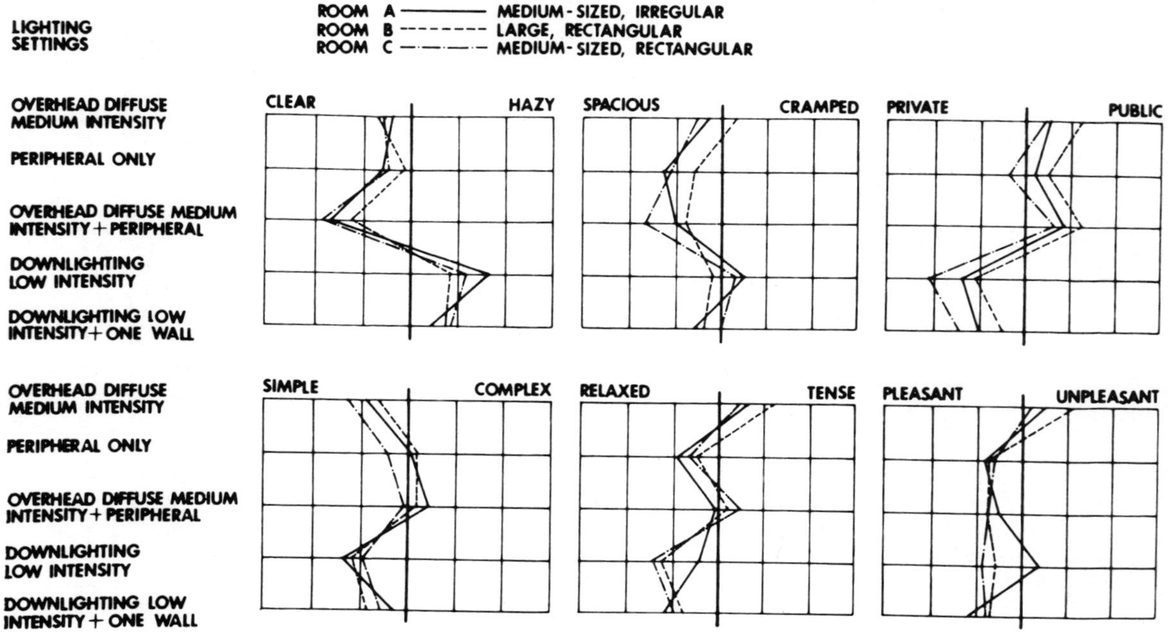

FIG. 5. The mean ratings for five different lighting installation types in three conference rooms on six rating scales. The mean ratings for each room on each scale are connected. The degree of similarity in profile for the three different rooms is good.[16] (Courtesy Illuminating Engineering Research Institute, New York.)

Will the same dimensions be relevant for say a factory or a shop, and even if they are, will the same lighting systems be assessed similarly in different contexts? A partial answer to this question can be obtained from Flynn *et al.*[16] who also used an auditorium in their experiment. This auditorium was assessed by two separate groups of people at different times. The first group saw seven lighting conditions and used a questionnaire made up of 13 rating scales. The second group had a questionnaire with 15 rating scales and saw 13 lighting conditions. The results for the five installations seen by both groups are presented on five representative rating scales, these being clear/hazy, public/private, unpleasant/pleasant, spacious/cramped and simple/complex. There is considerable consistency between the two groups of people in the rating of the installations on each scale. Further the relationship of the individual lighting installations on each scale is as before; greater clarity is related to higher illuminance, greater privacy is related to low illuminance from incandescent sources and peripheral lighting, greater pleasantness is produced by peripheral lighting and low illuminances, more spaciousness is related to the use of peripheral lighting and high illuminances, and a greater complexity is associated with the use of general lighting and wall washing. A comparison of these findings with those illustrated in Figs. 3 and 5 will demonstrate the similarity of the results. It should be emphasised that these various ratings have been obtained with different groups of subjects in a room with a different context, furnished in a different way. These findings do suggest some stability in the assessment of lighting. This conclusion is supported by the work of other people.

Hawkes *et al.*[17] have used factor analysis to examine people's impressions of a small rectangular office lit in different ways. The data collected is too extensive to be detailed here but an important condition was that the illuminance on the desks where the subjects sat was similar in all lighting conditions (approx. 500 lx). This experiment therefore tells us something about background lighting although both general and local lighting systems were used to light the desks. Eighteen different lighting situations were used made up of various combinations of conventional recessed luminaires, wall lighting, downlighting, desk lights and spot lights (Table 1). From the factor analysis of the ratings, excluding the evaluative scales, two dimensions were identified. One was a brightness dimension which had rating scales of 'dim/bright', 'strong/weak', and 'clear/hazy' related to it although it also had the 'cramped/spacious' scale. The brightness dimension relates fairly clearly to the overall illuminance in the room. The other dimension was simply called interest. The scales strongly

TABLE 1
INSTALLATION USED BY HAWKES ET AL.[17]

Installation number	*Installation*
1	Regular array of ceiling recessed fluorescent units with opal diffusers.
2	Incandescent downlights in regular array plus fluorescent wall washing of two end walls.
3	Regular array of ceiling recessed fluorescent units with bats-wing prismatic panels.
4	Fluorescent wall washing of two side walls.
5	Fluorescent desk lights at either side of each desk.
6	Incandescent spot lights at end of room and on desk.
7	Incandescent downlights in a regular array.
8	Incandescent spot lighting of side walls plus fluorescent wall washing of R.H. wall.
9	Regular array of ceiling recessed fluorescent units with prismatic panels.
10	Fluorescent desk lights at either side of desk plus fluorescent wall washing of L.H. wall.
11	Regular array of incandescent downlights plus incandescent spot lighting of two side walls.
12	Regular array of ceiling recessed fluorescent units with specular louvres plus fluorescent wall washing on R.H. wall.
13	Fluorescent wall washing of R.H. wall.
14	Regular array of ceiling recessed fluorescent units with specular louvres plus incandescent spot lighting of two side walls.
15	Incandescent spot lighting of all walls and desks.
16	Regular array of ceiling recessed fluorescent units with specular louvres.
17	Fluorescent wall washing of all four walls.
18	Fluorescent desk lights at either side of each desk plus incandescent spot lighting of two side walls.

related to this dimension were simple/complex, mysterious/obvious, uninteresting/interesting and commonplace/special. The distinction between installations here arises because those using diffuse sources were considered uninteresting but those using focused sources were not. The results for the individual installations (Table 1) are elegantly presented in terms of the two dimensions with iso-preference contours superimposed (Fig. 6). A study of Fig. 6 and Table 1 is well worthwhile. The preference ratings show a simple situation with the top righthand corner being the most preferred. Thus the most preferred background lighting is both bright

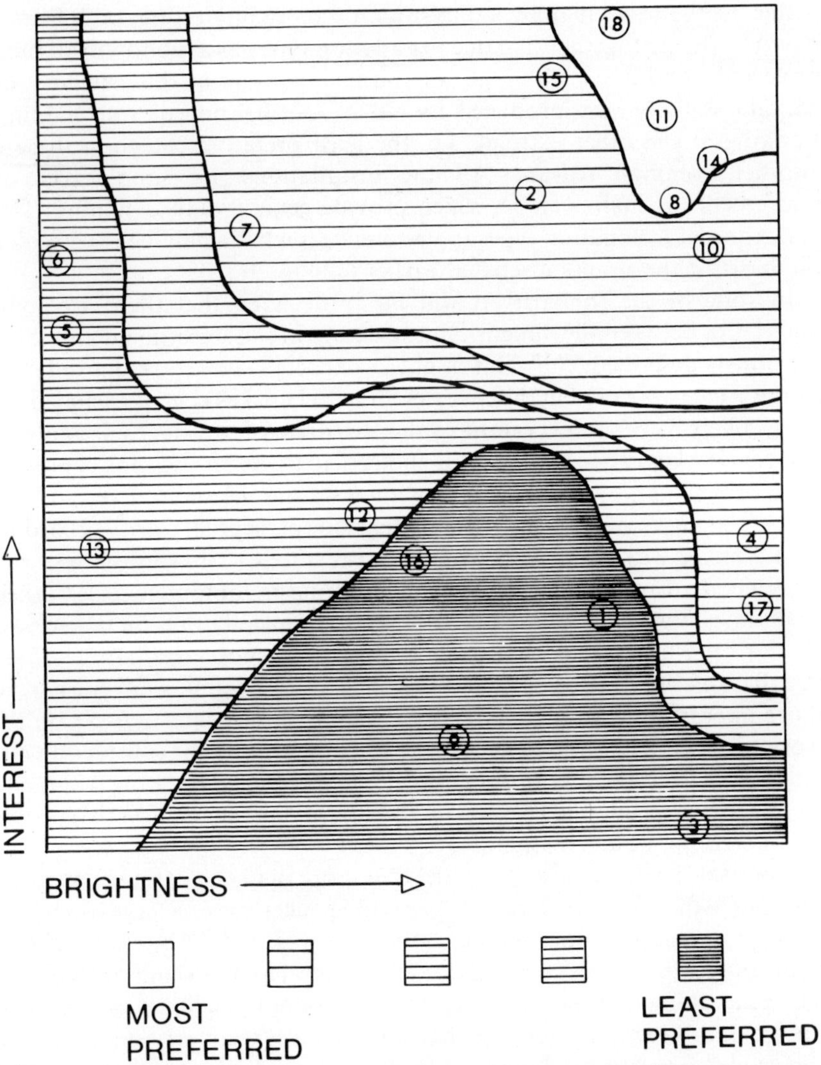

FIG. 6. A map showing the location of the types of office lighting listed in Table 1 on the two dimensions, interest and brightness, identified by factor analysis (17). Superimposed on this map are iso-preference contours which are based on preference ratings of the same installations. (Courtesy Illuminating Engineering Research Institute, New York.)

and interesting. Making an uninteresting situation brighter will not make it better, neither does making a dim situation more interesting, both factors have to operate together. No one installation type has a monopoly of most preferred conditions, although all the installations in this category do contain some variety produced by either spot-lighting or wall lighting. Looking at the other extreme, i.e. the least preferred category, there is another common feature. All the installations are simple recessed fluorescent luminaires alone, which provide general diffuse lighting. This allocation of installation types to preference is what would be expected on the basis of the results discussed earlier (cf Fig. 3).

In spite of the similarity it will be appreciated that there are some differences between the dimensional structure in this experiment and those previously described.[11,16] This might be expected since: (a) some different rating scales were used in the experiments, (b) Hawkes *et al.*[17] excluded the scales likely to be loaded on the evaluative dimension from their analysis, and (c) Hawkes *et al.*[17] asked their subjects to assess the lighting condition whilst Flynn *et al.* were concerned with assessing the room. Even so both sets of data agree on a bright/dim dimension (perceptual clarity according to Flynn *et al.*).[11,16] Also although the interest dimension was not identified by Flynn *et al.*, there is an interesting/monotonous scale in their evaluative dimension[11] and the distribution of installations on this scale is similar to what would be expected from the above results, i.e. those installations which use focused sources and/or wall lighting are considered more interesting than those which use merely diffuse light sources on the work surface. This degree of agreement is encouraging considering the differences in context, type of lighting systems examined, rating scales and people, between the studies. It certainly supports the suggestion that people assess lighting in a consistent way.

However, it should be appreciated that more than one lighting condition can lead to the same impression. Flynn and Spencer[18] used a large room with no specific context, other than a room to be assessed, lit by four different light sources in various combinations for general lighting and wall lighting. The general lighting was produced from a luminous ceiling which ensured diffuse non-directional lighting. The illuminances were very similar for all the eight conditions examined. The main difference between the installations was the colour properties of the light sources. Interestingly enough when the ratings of the interiors were used in a factor analysis four familiar dimensions were identified; the usual evaluative dimension, a perceptual clarity dimension, to which the scales 'bright/dim', 'hazy/clear', etc. were related, a spaciousness dimension and something the authors called 'complexity/novelty'. The first three of these dimensions are like the ones identified in the earlier experiments.[11,16] This supports the idea of stable dimensions on which lighting is assessed. However this time it is the lamp colour properties which sort the interiors along the dimensions. In particular the Cool White fluorescent lamp is rated as making the interior clear and distinct. By comparison the high pressure sodium lamp produces an interior rated as hazy and vague, a result not unexpected in view of the visual clarity results obtained by other methods.[19] Another interesting result is the fact that both the fluorescent Warm White lit interior and the Cool White lit interior are rated as more pleasant than the HID lit interiors and that both are even more pleasant when they provide both general lighting and wall lighting rather than general lighting alone. In fact the installations which have wall lighting produce ratings of greater clarity, spaciousness and pleasantness than those that do not. Although these results are interesting and consistent with results described earlier, it is important to note that the expectation that a separate colour dimension would be produced is not fulfilled. The colour scales are most strongly related to the complexity/novelty dimension, perhaps because the most obviously non-white source, the high pressure sodium lamp, is also a novelty in interior lighting of this type. In any case the failure of lamp colour to produce an independent dimension indicates its relative lack of importance for the interior examined. It should be noted that this result was obtained in a room where the decor had been deliberately kept to very low chroma colours. It is quite likely that interior colourfulness would be important in other circumstances.

It is also interesting to consider the importance of context when using the MDS method. Previously it has been shown that multi-dimensional scaling (MDS) gave good agreement with the factor analysis approach. Because the MDS procedure has also been tested over a number of contexts it too can be used to examine the stability of the dimensions obtained. Flynn *et al.*[16] used the auditorium mentioned earlier with six lighting installations similar in type to those shown in Fig. 2 to examine the repeatability of the three-dimensional structure found for the conference room. In fact seven installations were used, the seventh being high intensity downlighting. From the results in the conference room it was predicted that this installation should be considered as bright, overhead and non-uniform. It was found that observers used the same three dimensions as in the conference room in making their judgements of differences between the installations in the auditorium. The positions of the six repeated installations in the three-dimensional space were similar to those found before and the position of the high intensity downlighting installation was as predicted. There is thus some support for the view that the basic dimensions identified by the MDS method do represent consistent judgements.

Nonetheless these dimensions are not the complete story for all lighting situations. Hawkes[20] in an early study used the MDS procedure for examining lecture theatres. He established that four dimensions were necessary to explain the difference ratings but was not able to name the dimensions from the distribution of the lecture theatres in the space. Stone *et al.*[21] have also used an MDS procedure for lecture theatres but they only found three dimensions were required. Even Flynn *et al.*[16] have had some difficulty in reproducing their three-dimensional structure in their auditorium when other types of lighting system have been used. Obviously much more information is required before the variation of the MDS dimensions with context is understood.

By now what was a clear picture has become rather confused. Although the stability of the MDS dimensions is uncertain, there is evidence from the factor analysis studies of some consistency in the dimensions on which lighting is judged, and in the location of specific installation types on particular scales. This consistency supports Flynn's view that lighting provides a series of cues which people use to interpret a space and that these cues are somewhat independent of the room used.[22] However the evidence is limited to a small range of contexts and lighting conditions. There is also the complicating factor that different sets of cues can apparently lead to the same impression of an interior. It seems likely that the first part of Flynn's view is correct but until a much wider range of interiors and lighting conditions have been examined the independence from specific interiors, i.e. the generality of the cues, must be open to question. Table 2 gives a summary from Flynn[22] of the lighting cues he has identified and the impressions they produce. The usefulness of something like Table 2 to a lighting designer and the desirability of testing its generality should be obvious.

Although the information in Table 2 could be useful it is not necessarily enough unless the designer is sufficiently skilled to develop these simple ideas into hardware. What is really required is a method of linking the results of the subjective judgement of the lighting to the physical variables that the lighting engineer is experienced in handling. Hawkes[20] in his lecture theatre study achieved this. By taking multiple regressions between the MDS results and a mixture of physical variables such as horizontal illuminance, vector/scalar ratio etc., he was able to correlate each of his four

TABLE 2
LIGHTING CUES FOR SUBJECTIVE IMPRESSIONS[22]

Subjective impression	*Lighting cues*
Impression of perceptual clarity	Bright, uniform lighting. Some peripheral emphasis, such as with high reflectance walls or wall lighting.
Impression of spaciousness	Uniform, peripheral (wall) lighting. Brightness is a reinforcing factor, but not a decisive one.
Impression of relaxation	Non-uniform lighting. Peripheral (wall) emphasis, rather than overhead lighting.
Impression of privacy	Non-uniform lighting. Tendency toward low light intensities in the immediate locale of the user, with higher brightnesses remote from the user. Peripheral (wall) emphasis is a reinforcing factor, but not a decisive one.
Impression of pleasantness	Non-uniform lighting. Peripheral (wall) emphasis.

dimensions strongly with some combination of the physical quantities. Stone *et al.*[21] have been able to achieve the same thing, although the physical variables were rather different.

This opens up a useful route for moving from the observer's assessment of an interior to the physical variables required to influence that perception. By means of an MDS analysis the number and sometimes the identity of the dimensions involved in the observer's assessments can be established. If necessary, factor analysis can apparently be used to name the dimensions. These dimensions can also be related to the physical characteristics of the interior. This is what the designer wants since it is the physical characteristics that he has to use but it is the observer's perception of the interior that he is trying to influence. It is the realisation that this route exists that makes the use of factor analysis and MDS in specific installations interesting. By comparison the simple correlation method used in the past is limited, potentially misleading and uninformative. To try to develop it further would be to attempt to improve the bow and arrow in the age of the guided missile.

Finally in this section, a word of warning. In all experiments in which the aim is to obtain information on the impression an interior gives, there is a tendency to consider the conditions as a picture with no relevance to everyday living. This is a tendency which should be strongly resisted. To do this it is necessary to establish a particular context for the interior being assessed, and to use representative observers who are familiar with the particular situation. The importance of identifying the group of people from whom information is required is illustrated by a result reported by Ne'eman.[23] He found that sunlight in hospital wards was regarded as very pleasant by the patients but as a nuisance by the nurses. Thus different groups of users of an interior may expect different things from the lighting. As an ideal, office lighting, for example, should be assessed by office workers who know what is required from lighting in an office. This ideal situation is not always possible but careful experimental control can ensure it is approached. Putting observers into a real room will be better than showing them pictures of a room; giving the observers some of the tasks that are usually done in the room will also establish the context, as will a careful description of the function of the room in the observer's instructions. The problem is essentially one of creating realism but at the same time leaving enough flexibility so that the experiment is still a practical possibility.

This raises the other problem with the MDS/factor analysis approaches. The previous discussion has demonstrated that these analysis methods are much superior to the earlier methods in the quality of information they produce. However they are equally superior in the resources required to generate the information. Ideally a number of real installations in rooms of the same function assessed by a representative sample of people are required. For the MDS techniques all the possible pairs of rooms should be seen by all the observers. In the factor analysis method each observer has to see all the installations. The logistics of such exercises are obviously complex and expensive. It therefore seems likely that in spite of their obvious advantages the MDS/factor analysis approaches will only be occasionally used unless there is some way of simplifying these logistics. One obvious approach is not to use a real interior but to use colour slides, or colour film. Shafer and Richards[24] have tried this approach for outdoor scenery and Hershberger and Cass[25] have used it for housing and commercial buildings. Comparisons between people's rating scale assessments of the real scene and colour slides of the scene show strong similarities, as do the underlying factor structures. This suggests that slides which depict most of the variation of the environment can be used as substitutes for the real thing, although the danger of considering the lighting as part of a picture rather than real life is obviously enhanced. Flynn *et al.*[16] have examined the same possibility for interiors. Slides were made of the six installations in the conference room (Fig. 2). Factor analyses were done for ratings of the room by people sitting in it and for the ratings of the room by people looking at the photographs. The dimensions identified by the two methods were almost identical. In addition the overall pattern of mean ratings for the six installations was similar for both methods although there were differences over specific scales. The conclusion is that carefully taken slides can give a reasonable overall picture of the dimensions involved but may lead to some erroneous conclusions about particular installations on particular scales. Obviously there is no substitute for the real interior presented to a representative group in a realistic way, but it can be argued that with care in checking some conditions against the real interior, it would be justifiable to use slides. Certainly the potential benefits in making the MDS/factor analysis approaches widely available are too great for the photographic substitute to be rejected out of hand.

DEEDS NOT WORDS

All the work described previously has been concerned with allowing people to state their impressions of a lit interior in some way. This can tell us a great deal about their opinions but it is essentially a monocular view of the influence of lighting. For a binocular view it is necessary to consider behaviour. Does lighting influence behaviour and if so, in what way? It is well known that lighting affects the performance of tasks but that is not the sort of behaviour of interest here. What is of interest is the way in which different lighting conditions can cause changes in the way an interior is used. To study this question the experimenter uses simple observation and recording, usually covert observation and recording. The idea is to put people into a situation in which they are unaware of the observer and then to monitor their responses and comments. This has similarities with the methods of the zoologist observing the behaviour of an exotic animal in its habitat. In this case the animal is man, the habitat is the luminous environment. The zoologist analogy does not stop at the method of data collection. It also applies to the analysis of the data. Advanced statistical methods are rarely used—the analysis is usually intuitive and often the results are open to several interpretations. In a sense, such behavioural studies can be regarded as a backward step in precision from the questionnaire studies discussed previously but they can be justified as a forward step in realism and in some cases relevance. Patterson and Passini[26] have summarised the advantages and disadvantages of behavioural studies and questionnaire studies for examining physical conditions.

Unlike the studies discussed previously, behavioural work which has involved lighting as a variable forms a very *ad hoc* collection. Taylor and Sucov[27] considered the effect of the lighting of passages on the choice of which passage to use, and showed that if faced with two passages, the more brightly lit one would be used by most people. In fact this study could well serve as a warning of the difficulties of behavioural work. Sanders *et al.*[28] measured the effect of illuminance level and distribution on the noise level of groups of people talking whilst waiting in an assembly room. They found that a low, uneven illuminance pattern tended to be associated with lower noise levels than a higher, more even illuminance pattern. It would be interesting to know to what extent this finding depended on the context. Flynn *et al.*[11] had a coffee bar lit in two different ways. The seating arrangements chosen by the subjects were noted (Fig. 7). The upper diagram shows that in the first lighting condition (A) when the subjects approached the unoccupied tables they chose to look towards the area of the room lit by wall lighting in a way that had been rated as interesting and pleasant. Although they looked towards this area, the subjects actually chose to sit in an unlit area. Possibly the perils of coffee spillage were considered more important than the benefit of better lighting. The lower diagram (B) shows what happened when the lighting of the coffee bar was changed. In the second installation the apparent room size was halved and one wall was lit. In this situation the subjects still chose to sit near the coffee bar in an unlit area but they now chose seats that enabled them to look towards the lit areas. Obviously lighting does influence behaviour but it is only one factor amongst many.

One final example will serve to illustrate the understanding that can be achieved by simple observation. Flynn *et al.*[11] recorded the observers' behaviour and remarks to each other on entering the conference room lit as described in Fig. 2. It is interesting to compare these observations with the quantitative results obtained after much computation. Installation 3, which was diffuse overhead lighting of low illuminance, was greeted with remarks such as 'reminds me of an elevator' and thumbs down gestures. Similarly installation 5, which was a high illuminance diffuse overhead lighting, caused rubbing of the eyes and remarks like 'we have ways of making you talk'. Both these installations were considered unpleasant on the evaluative dimension in the factor analysis as would be expected. Thus the evaluative dimension is available from the simple observation. However installation 5 was also rated as giving good perceptual clarity and a moderate degree of

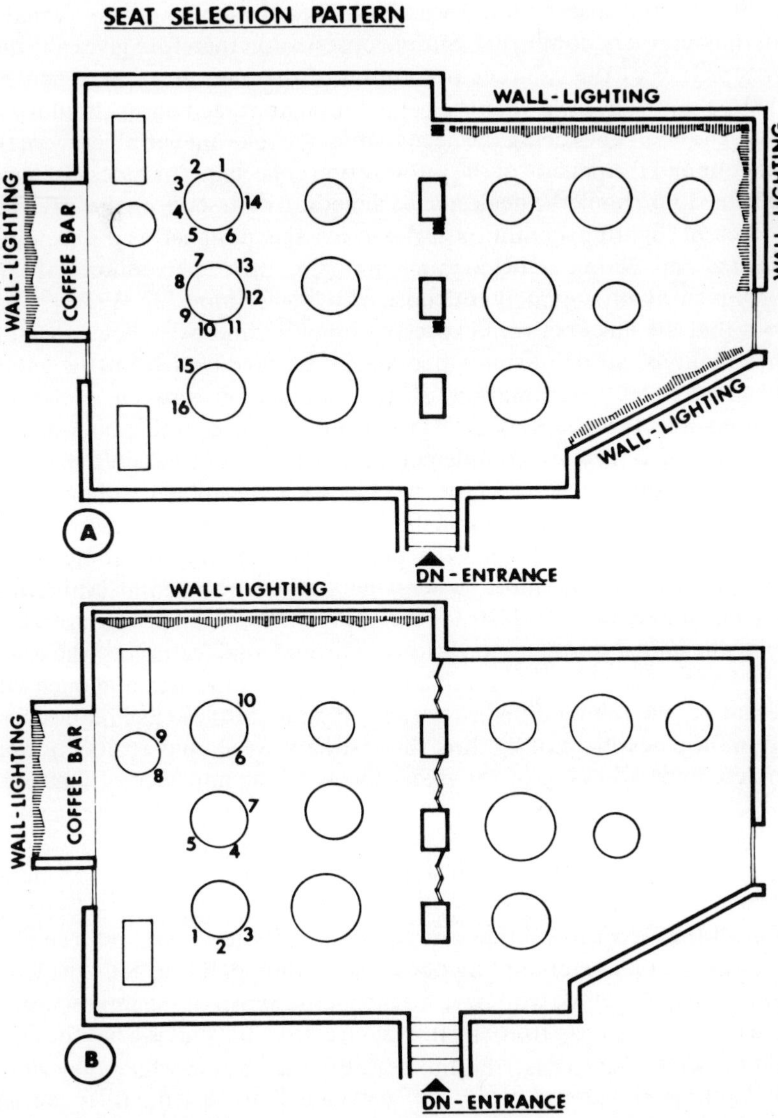

FIG. 7. Seating plans for the two lighting installations in the coffee bar.[11] The numbers give the position and order of seating chosen. (Reprinted from the October 1973 issue of *Journal of the Illuminating Engineering Society* with permission of the Illuminating Engineering Society of North America.)

spaciousness but installation 3 was not. These insights are not available from the observers' comments. Simple observation therefore gives a rather coarse picture of the subject's perception of the interior.

It should now be clear that the behavioural approach has its limitations as well, the main ones being the need to infer the relevant variables from the behaviour and the paucity of the information collected. This does not mean that behaviour should be neglected as a method of describing the influence of different lighting conditions. After all the designer is often more interested in people's behaviour than in their perception of the environment. Unfortunately for those who like a simple life Wicker[29] has shown that the links between expressed opinions and overt behaviour are rather tenuous, so one cannot necessarily be predicted from the other. Thus a complete examination of the interaction between a lighting condition and the people who use it would involve both opinions and behaviour since they are complementary. If this is not possible then it is necessary to consider which of the two aspects of the influence of lighting is of greater importance to the specific situation being considered.

For the future there is also the possibility of taking up some of the techniques used in the more general field of environmental evaluation. These techniques include the trade-off approach in which people are asked to allocate limited resources to various environmental features,[30] the use of model kits to determine preferred conditions[31] and the idea of games with different people taking different roles.[32] If some of these techniques were tried in the assessment of lighting then perhaps we should achieve a more complete understanding of the many facets of the influence of lighting.

WHAT ABOUT THE ARTIST?

But is all this experimentation and analysis necessary? After all exciting, interesting and pleasant lighting does exist, although it is usually the work of gifted people, people who have an instinctive grasp of the impression to be created and the ways to create it. If such artists, for that is what they are, could be persuaded to describe and quantify the ways in which they work, and lighting designers could be persuaded to listen, then maybe investigations of the type described here would not be needed. Voltelen[33] discusses the advantages of listening to artists most vividly. But this seems to be wishing for the moon. There is little contact between such artists and lighting designers. The aim of the approaches discussed is to provide alternative routes to an understanding of the way in which lighting

FIG. 8. The Los Cerritos Shopping Center with its river of light. (Reprinted from the June 1976 issue of Lighting Design and Application with permission of the Illuminating Engineering Society of North America.)

influences people. The one advantage the scientific approach has over the artistic is that the advice it generates is firmly rooted in a large number of people's responses to the lighting. The artist is unique in many ways, one of which may be his responses to lighting. In spite of this apparent advantage any scientific method is ultimately dependent on the artist. Nothing can be learned about a lighting installation that is not examined. The artist is the creator of lighting effects, the scientific method can only analyse the finished product. Possibly when the process has been done enough times to generate guidelines to the ways in which various lighting features influence people, then the guidelines could give new inspiration to the artist. This stage is a long way ahead. At the moment we are still dependent on the artist. There is no better way to end the chapter than to demonstrate his skill. Figures 8, 9

FIG. 9. A striking reception area. (Courtesy Yorkshire Electricity Board.)

FIG. 10. A private office with a combination of daylight and electric light.

and 10 show some of the lighting effects that are possible in everyday installations. When we have a general understanding of the features that influence the impression created and how to apply the underlying principles in different contexts then we shall indeed have moved beyond our present stage of avoiding discomfort. Vision is the most stimulating of the senses in the subtlety and complexity of its effects. It is a great instrument. Let us learn to play it.

REFERENCES

1. Illuminating Engineering Society, 'IES Code for Interior Lighting', 1977.
2. SAUNDERS, J. E., *Ltg Res. Technol.*, 1969, **1,** 37.
3. HOPKINSON, R. G., *Architectural Physics, Lighting*, 1963, HMSO London.
4. INUI, M. and MIYATA, M., *Ltg Res. Technol.*, 1973, **5,** 103.
5. HEWITT, H., KAY, J., LONGMORE, J. and ROWLANDS, E., *Trans. Illum. Engng Soc.*, (*London*), 1967, **32,** 63.
6. HAWKES, R. J., *Building*, 1970, 19th June, 69.
7. JAY, P. A., *Ltg Res. Technol.*, 1971, **3,** 133.
8. LYNES, J. A., *Ltg Res. Technol.*, 1971, **3,** 24.
9. KIMMEL, P. S. and BLASDEL, H. G., *J.I.E.S.*, 1973, **2,** 113.
10. LYNES, J. A., *Ltg Res. Technol.*, 1977, **9,** 51.
11. FLYNN, J. E., SPENCER, T. J., MARTYNIUCK, O. & HENDRICK, C. *J.I.E.S.*, 1973, **3,** 87.
12. CANTER, D., *Psychology for Architects*, 1974, Applied Science, London.
13. HARMAN, H. M., *Modern Factor Analysis*, 1970, University of Chicago Press, Chicago.
14. KULLER, R., 'A Semantic Model for Describing the Perceived Environment', National Swedish Building Research D12, 1972.
15. KRUSKAL, J., *Psychometrika* 1964, **29,** 1 and 115.
16. FLYNN, J. E., SPENCER, T. J., MARTYNIUCK, O. and HENDRICK, C., Interim Report to Illuminating Engineering Research Institute, New York, Project 92, 1975.
17. HAWKES, R. J., LOE, D. L. and ROWLANDS, E., 'A Note Towards the Understanding of Lighting Quality', Report to the Illuminating Engineering Research Institute, New York. Project No. 96–71, 1977.
18. FLYNN, J. E. and SPENCER, T. J., *J.I.E.S.*, 1977, **6,** 167.
19. BOYCE, P. R., *Ltg Res. Technol.*, 1977, **9,** 11.
20. HAWKES, R. J., 'A Study of Lighting Quality.' Report to the Illuminating Engineering Research Institute, New York, Project 96–70, 1970.
21. STONE, P. T., PARSONS, K. C. and HARKER, S. D. P., *Ltg Res. Technol.*, 1975, **7,** 259.
22. FLYNN, J. E., *Ltg Des. App.*, 1977, February, 6.
23. NE'EMAN, E., *Ltg Res. Technol.*, 1974, **6,** 159.
24. SHAFER, E. L. and RICHARDS, T. A., in *Psychology and the Built Environment*, (ed. D. Canter and T. Lee), 1974, Architectural Press, London.

25. Hershberger, R. and Cass, R. C., EDRA5, *Man-Environment Interactions: Evaluations and Applications*, (ed. D. Carson), 1975, Dowden, Hutchinson and Ross Inc., Stroudsburg, USA.
26. Patterson, A. H. and Passini, R., EDRA5, *Man-Environment Interactions: Evaluations and Applications*, (ed. D. Carson), 1975, Dowden, Hutchinson and Ross Inc. Stroudsburg, USA.
27. Taylor, L. H. and Sucov, E. W., *J.I.E.S.*, 1974, **3,** 237.
28. Sanders, M., Gustanski, J. and Lawton, H., *J. Appl. Psychol.*, 1974, **4,** 527.
29. Wicker, A. W., *Journal of Social Issues*, 1969, **25,** 41.
30. Banerjee, T. K., Baer, W. C. and Robinson, I. M., EDRA5, *Man-Environment Interactions: Evaluations and Applications*, (ed. D. Carson), 1975, Dowden, Hutchinson and Ross, Inc., Stroudsburg, USA.
31. Rivlin, L. G., Rothenburg, M., Justa, F., Wallis, A. and Wheeler, F. G. EDRA5, *Man-Environment Interactions: Evaluations and Applications*, (ed. D. Carson), 1975, Dowden, Hutchinson and Ross Inc., Stroudsburg, USA.
32. Mackie, D., *Architectural Psychology*, (ed. R. Kuller), 1973, Dowden, Hutchinson and Ross Inc., Stroudsburg, USA.
33. Voltelen, M., *Architectural Psychology*, (ed. R. Kuller), 1973, Dowden, Hutchinson and Ross Inc., Stroudsburg, USA.

Chapter 4

LIGHTING AND THE SEARCH FOR MEANING

P. A. JAY
M.A., F.C.I.B.S., F.I.O.A., M.Inst.P., M.Cons.E.
Senior Partner, Peter Jay and Partners, Consulting Engineers, London, UK

SUMMARY

There are two paradoxes well known to lighting engineers. The first is between common experience and the implications of purely psycho-physical studies of limited aspects of the visual process. The second is the basic difference between current practice for the lighting of working areas and that for display and drama.

It is shown that both conflicts are largely resolved if modern ideas of the psychology of visual perception are studied. In particular, it is shown to be helpful to regard visual perception as Information Processing, and to re-interpret many psycho-physical experiments in terms of Information Theory.

Studies of the visual process quoted in textbooks or papers concerned with lighting often give the impression that perceiving objects clearly and accurately is in some way very difficult and that it is essential that the lighting shall always be most carefully arranged if we are to see satisfactorily. However, in most cases we can see well enough even when the lighting breaks all the rules of good practice. This is not to suggest that lighting engineering is of no value, and there is a big difference between being able to see adequately, as we generally can, and having the lighting so arranged that we can achieve high visual performance for many hours at a time with minimum effort. Even so, there is apparently a paradox; study of psycho-physics, threshold contrast, colour discrimination and so forth appear to indicate that vision should be highly dependent on the luminance pattern and since, for example, the luminance pattern in a sidelit room will be quite different from that obtained from a regular arrangement of luminaires on the ceiling, we should expect our perception of the room to be quite different in the two cases.

However, our perceptions are likely to be much the same in practice, and differences only appear if some major mistake has been made. Sources of poor colour rendering can distort certain pastel shades, we may suffer glare either from the sun seen through the window, or from the electric sources; if the overhead lighting is purely by downlighters the walls may then appear disagreeably dark and the room become gloomy, but even in cases such as this very large changes in luminance pattern and spectral composition have only a minor effect on our perception of the room and the objects in it.

Most people if asked to comment on the difference between the effect of side lighting and top lighting would simply say that one is the effect found by day and the other by night. If we pressed them to describe the differences in greater detail they might come up with some of the points listed above, but they would have to think about it rather carefully. In fact, as appraisals have shown, it is very difficult to get intelligent comments on lighting from people who have had no practice or training in the subject.

To put it another way, of perhaps more importance to those who specialise in lighting: it is apparently rather easy for semi-skilled people to design a lighting installation which is merely adequate. Probably most installations arise in this way, employing luminaires selected and sited by people with very little knowledge. The annoying thing to us specialists is that such installations are generally quite acceptable and complaints arise only where some obvious mistake has been made. Even then it is often quite easy to put right.

How can we reconcile our knowledge that the luminance pattern is critically dependent on the lighting with the robustness of our ability to perceive our surroundings accurately? Let us restate the problem: the stimuli to our eyes can be described by physicists as luminance patterns which are continually changing. However, our perceptions are of a stable world consisting of solid objects whose relative positions may change but whose appearance remains relatively stable, and independent of our movements or of the direction from which we look at them. Further, when viewing our surroundings we may often be aware that the lighting is changing, but this very rarely results in a change in the appearance of objects themselves.

A visual system which did not perceive a stable world in the manner described would not be of much survival value, since in a natural environment the lighting varies enormously both by day and night not merely in intensity but in colour, direction, degree of diffusion and so forth.

Does it really matter to lighting engineers whether we can explain the paradox? In fact it matters very much, since if the kind of information which psycho-physicists obtain in laboratory studies is really relevant to perception, certain conclusions follow about the way we should design lighting installations and the amount of light we should provide. If, however, not all the information we obtain from laboratory studies is directly relevant to perception in practice it may well be that rather different conclusions about what sort of lighting we need for various purposes may follow. We may even manage to say something constructive about another paradox which has bothered the lighting profession for a long time. Certain kinds of research have indicated that quite high illuminances are required for common tasks such as reading and checking columns of figures, but current practice in many countries indicates that such illuminances are not always necessary. Although most experienced lighting practitioners generally manage to make sensible decisions in any given case, this contradiction between theory and practice is always rather worrying and often induces neurosis in those who have to draft Codes and Standards, since we would all like to feel that our recommendations are soundly based and will withstand detailed critical examination. This cannot be the case when theory and practice are widely at variance.

THE TRADITION OF PSYCHO-PHYSICAL RESEARCH

Most of the techniques of psycho-physical research evolved in the 19th century, and were intended to find out how the sensory processes actually work. Their application to practical engineering situations is more recent and the justification has rarely been seriously examined. Further, the underlying theory of much 19th century psychology was borrowed from the English Empirical Philosophers of the 17th and 18th centuries, especially John Locke, who asserted that all knowledge must come from experience, which can itself only be obtained through our sense organs. At that time physical science was starting to emerge in a recognisable form and the immense success of physical concepts led people to suppose that everything could ultimately be explained in terms of the Laws of Physics. This may well be true, but our understanding of the laws of physics changes and develops with time and it by no means follows that physical optics, as understood for instance by Newton, is necessarily an adequate basis for the elucidation of problems of visual perception.

Early studies of our sense organs implicitly assumed that they are physical measuring instruments and described their performance in such terms. Only a few years ago it used to be said that 'the eye is a very bad photometer' until J. M. Waldram pointed out that it is not a photometer at all but the organ of sight. Even so, if the eye is not a photometer, that is to say, if it is not measuring light, what is it measuring? It certainly responds to light and at an elementary level it does have a few of the characteristics of a photometer, in the sense that in the absence of light it gives no ordered response but only 'noise', while when stimulated by light its response does to some extent increase with the strength of the stimulus. There is a vast literature on the absolute and differential thresholds of the eye, while much ink has been spilt on the question of whether its response at supra-threshold follows a logarithmic expression or a power law. However all this data has been obtained under controlled conditions in the laboratory and, although so often quoted in the literature, its relevance to the functioning of the visual system in more normal circumstances is at first sight rather obscure.

Psycho-physicists are, of course, fully aware of the discrepancy between the sort of things they measure and visual perception in practice but, following John Locke, often explained the discrepancy in terms of learning in infancy, resulting in a process of 'unconscious inference' in the adult. As is generally known, Helmholtz used this phrase to explain why it is that a piece of coal in sunshine still looks black although its luminance may be higher than that of a piece of white paper in shade. Such an explanation seems to imply that we cannot perceive things accurately unless we already know what they are—but, as explained above, a visual system which can do no better than this would be of rather limited survival value.

THE VISUAL SYSTEM

Every age is limited by the level of complexity it can conceive, and this in turn is related to its technological attainments. Berkeley in the 18th century referred to the image 'painted on the retina': in the 19th and 20th centuries we use instead the analogy between the eye and the camera, but we can now if we so wish think of cine and television cameras instead of still cameras. Similarly, late 19th and early 20th century writers would often draw an analogy between the brain and a telephone exchange, but we now think of it as an *information processing system* which at least means that the range of analogies available to us in discussing perceptual problems has been enormously increased.

Further, our increasing knowledge of systems engineering has helped us to understand that when studying the functioning of a complete system we must take into account its purpose, since it may have properties that could not easily, if at all, be deduced from an analysis of its component parts. This might suggest that study of the response of the eye to limited visual stimuli is not a very reliable guide to its normal operation. In fact, it turns out that some of the detailed studies of eye function are helpful in explaining certain features of the visual process taken as a whole, but the significance of these studies only becomes apparent when we realise that we are trying to explain the functioning of the system, and not merely of 'the eye' itself.

The function of the visual system is to enable us to perceive things which are important to us, and these are generally objects. Objects or parts of objects are bounded by contours and often differ in colour or texture. Within a specific contour differences in colour are generally less. These characteristics of objects will be largely independent of the ambient light, or of our direction of view, since although the spectral composition of the light reflected from a surface will depend not merely on its reflectance at each wavelength but also on the composition of the incident light, two different surfaces are likely to continue to reflect differently even when the composition of the light varies. Similarly, although the projected shape of an object on the retina will vary with our direction of view and the orientation of the object, the object will continue to be bounded by a contour of some kind. J. J. Gibson[1, 19] has developed a most elegant theory of visual perception based on what he calls 'Ecological Optics' and he describes a large number of 'stimulus invariants'—that is to say, features of the scene which will remain substantially the same, or subject only to distortion in an ordered way, as the observer or the objects move. However, it is not necessary to accept the whole of this theory to realise that a visual system, if it is to be of any use, must be particularly efficient in detecting contours and colour changes in the environment, and need be less sensitive to major variations in brightness of the whole scene, such as will occur from sunrise to sunset.

DETECTION OF CONTOUR

Laboratory studies have shown that stimulation of one part of the retina changes the sensitivity of other parts, and especially adjacent parts. The phenomena are variously called 'reaction', 'induction' and Mach Bands'. Laboratory data on highly artificial stimuli is both voluminous and confusing, but we should realise that there is no reason to expect that the interactions between different retinal stimuli should always be the same as if there were a fixed pattern of retinal interconnections, as in a telephone exchange. We now know enough to realise that there is a vast number of *potential* retinal interconnections and interactions which may be activated under different conditions of stimulation.

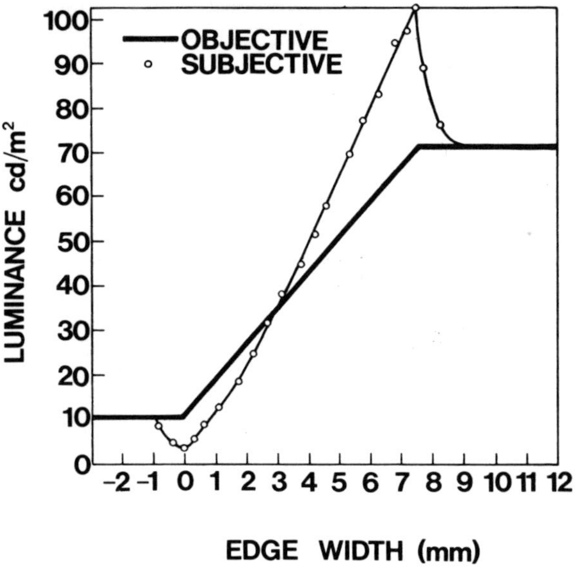

FIG. 1. The subjective brightness across a luminance gradient.[2]

Figure 1 is from Lowry and da Palma[2] and shows the 'apparent brightness' across a luminance gradient. This data has been quoted first, since it was obtained using a relatively large stimulus field under relatively 'realistic' conditions since the observers were required to adjust the luminance of a movable slit until it appeared as bright as the adjacent segment of the gradient. The results are quite consistent with many more carefully analytical studies of retinal induction, although not necessarily with all of them. As explained above, this should not surprise us.

If we prefer, we can explain such results in terms of pure adaptation. It is known that the eyes undergo three types of reflex movement, saccadic flicks over several degrees of arc, a continual oscillation over a few minutes of arc, and a slow drift between saccades. As the eyes scan a sharp luminance gradient or a contour receptors previously stimulated by a bright part of the field will subsequently be stimulated by a darker part, and vice versa. This will go some way to explain the results of Fig. 1, but it seems likely that neural interaction is also involved.

For instance, animal experiments[3] have found optic nerve fibres which are responsive to lines of given orientation, and this form of response appears to arise if not actually in the retina, at least quite close to it from the neurophysiological point of view.

The structure of the retina is extremely complex and the description still given in many textbooks, in which each separate foveal cone is connected more or less directly to an individual optic nerve fibre, is certainly untrue. Nineteenth century workers assumed that virtually all processing of the retinal image had to take place in the higher centres of the brain, since they lacked the detailed knowledge of neurophysiology and of systems engineering which we are now beginning to acquire. We can conceive of a retinal structure which responds to significant relationships between adjacent parts of the visual field and transmits to the optic nerve not merely data about little dots of 'brightness' but information about contours and gradients.

BRIGHTNESS CONSTANCY

The stimulus to each individual retinal receptor can be described in terms of luminance and spectral composition only. The luminance of the receptor will be a substantially constant fraction of that of the surface viewed. A given luminance can be achieved by illuminating a surface of any reflectance appropriately, and luminance of itself therefore gives us no information about the reflectance of a surface. If we close one eye and look at a *uniform* surface through a black tube (a reduction tube) we cannot differentiate a brightly lit grey surface from a dimly lit white surface, as anyone can confirm for himself. This does not necessarily work if the surface is in any way textured, but if conditions are carefully arranged we do, apparently, perceive largely in terms of luminance and cannot determine the *lightness* of the surface seen through the tube. However, this experiment is not all that easy to perform, and there is no question that under normal conditions, and when viewing a complete scene with all its contours and contrasts, we do not make such confusions and white paper in shade is nearly always clearly distinguishable from coal in sunshine. Is this due purely to learning in infancy, or should we look elsewhere for the primary explanation?

A great deal of laboratory work has been done on this kind of thing, and Fig. 2 is from Saunders.[4] The 'apparent brightness' seen in one eye was obtained by a haploscopic matching technique (i.e. by matching to a stimulus seen separately against a background of constant luminance by the other eye) which avoids some of the difficulties of direct numeration, another technique which has been extensively used.

The data has been drawn in such a way as to show the 'Apparent Munsell Value' of surfaces of different luminance seen against a background of reflectance 0·2. The reference conditions correspond to an illuminance of 786 lux, and under these conditions the Apparent Munsell Values have been normalised to the actual Munsell Value. It will be seen that variations in appearance over a range of illuminances of 10:1 or more are comparatively small. It is difficult to account for this by a process of learning in infancy since no identifiable objects were used in the experiment.

Again, it turns out that it is possible to construct an explanation of such effects as inherent properties of the visual process which have been studied

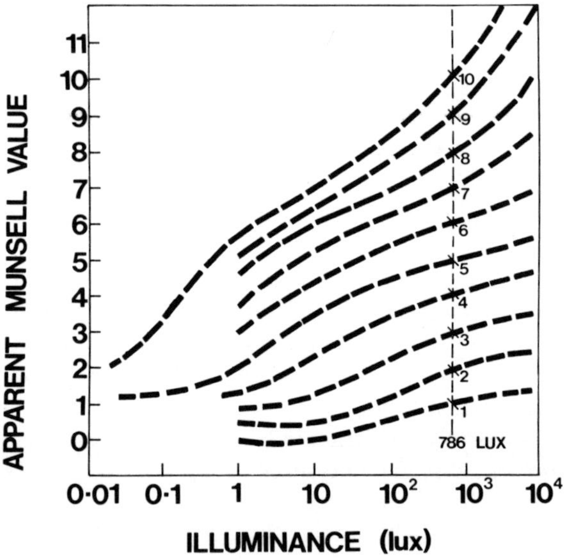

FIG. 2. The 'Apparent Munsell Value' of surfaces of different luminance seen against a background of reflectance 0·2. The vertical dotted line shows the reference condition, at an illuminance of 786 lux. In this case 'apparent' and actual Munsell Value are the same.[4]

in the laboratory, such as gross adaptation to ambient light, local adaptation of small parts of the retina, successive induction and spatial interaction.[5]

A MODEL OF THE VISUAL PROCESS

Jay[6] has shown that it is possible to construct a model of the visual process which accounts for its ability to extract the maximum amount of meaningful information from the visual scene, largely independently of lighting conditions. All such models must be tentative and unsatisfactory, and yet some kind of model is essential in discussing problems of this kind, since few people can accept as facts data which they find it quite impossible to understand. This was the attraction of the 'learning in infancy' school since it appeared to offer an explanation of phenomena which were, until at least the second quarter of the 20th century, totally incomprehensible. We can now see that learning in infancy does not in fact 'explain' anything at all—it merely provided an excuse for excluding certain kinds of phenomena from one's research. Nevertheless it did assist scientific advance in its day since it enabled psycho-physicists to start on a form of investigation which would otherwise have appeared so complex as not to be worth undertaking at all.

A similar justification can be advanced for the naive and unsatisfactory attempts to 'appraise' lighting installations in many countries. It is, however, to be hoped that nobody will attempt to apply such primitive and incomplete research results to practical situations generations before they are capable of doing so—as happened to late 19th century psycho-physical results.

CONDITIONS FOR ACCURATE PERCEPTION—1

Figure 2 shows that our ability to perceive accurately is largely independent of lighting conditions, but not absolutely so. For instance, where illuminance varies by much more than ten to one the changes in Apparent Munsell Value become more significant, while at high illuminances the AMV of light surfaces exceeds 10. This can only mean that they are no longer perceived as opaque surfaces, but take on some of the qualities of objects which emit or transmit light. It follows that perception is likely to be most accurate where contrast is fairly low, say, a ratio of maximum to minimum illuminance of about ten to one—certainly not more. Further, variations in illuminance should be relatively gradual: even under optimum conditions we are quite frequently deceived by sharp shadows, such as are cast by the sun or small electric sources, since the same processes which tend to sharpen contours in perception will also tend to sharpen the boundaries of a shadow, and if its shape and position are appropriate we may mistake the edge of a shadow for the boundary of an object, or between surfaces of differing lightness.

It should also be noted from Fig. 2 that as illuminance decreases the curves of AMV all draw closer together. This means that surfaces whose lightness differs by only a small amount are more difficult to discriminate. This again is in accordance with well-known psycho-physical data. Further, of course, photopic vision is not fully developed at low illuminances.

Conditions for accurate perception may be summarised as follows:

(a) Illuminance shall everywhere be high enough to permit photopic vision and the discrimination of all necessary detail. For normal scenes this implies that the minimum illuminance shall be above 10 lux, although not necessarily as much as 100 lux. It does not follow, of course, that we can carry out fine visual tasks for long periods at such illuminances without effort; however we are not in this case discussing a working situation, but merely those conditions necessary to perceive our surroundings satisfactorily.

(b) The range of illuminances simultaneously in view shall be not more than about ten to one and preferably less, while any variations in direct illuminance shall be gradual.

(c) The spectral composition of the light shall be such as to permit appropriate colour discrimination.

It can immediately be objected that only condition (b) is really meaningful and precise. In condition (a) what is 'all necessary detail'? Surely, this must be the 'critical detail' on which illuminance recommendations used to be based, on the Weston–Beutell Principle. This will be discussed more fully below. Again, what is 'appropriate' colour discrimination, and what range of spectral compositions permits it? This again will be discussed more fully in a later section.

THE CONCEPT OF CRITICAL DETAIL

The meaning of critical detail used to be so obvious. It was the detail we needed to perform a given visual task, and for much visual data could be operationally defined. When carrying out Landolt ring tests it was the size of the break in the ring, and in reading it was, for instance, the break in letter 'c' which enables us to distinguish it from letter 'o', etc., which comes to much the same thing. However, in Great Britain the Weston–Beutell approach, although very elegant, was not in the long run found to be an adequate basis for Lighting Codes, although heroic efforts are still being made to employ the same underlying idea in the US. However, despite the monumental work of Blackwell, or perhaps indeed because of it, it has been found impossible to predict visual performance from a study either of the apparent size and contrast of the 'critical detail' in a given task, or even of the threshold contrast as measured by a Visibility Meter. CIE Report 19, despite the formidable evidence there assembled, gets us no further in predicting what illuminance is required for a given task to achieve a given standard of performance without carrying out the task under controlled conditions in order to obtain the appropriate parameters to put in the equations. Unless and until we have some method which enables us at least to make a preliminary estimate of these parameters from mere inspection of the task itself, and without carrying out additional experiments, we cannot really be said to have any useful practical information on lighting in relation to visual performance.

What is the problem and what has gone wrong? Could it be that the whole idea of critical detail and threshold contrast is unsound? This seems unlikely since there are certain tasks, such as inspection, where the whole job does appear to consist in perceiving and responding to certain kinds of detail. In these cases we should expect laboratory data to provide a reasonable guide, although it is often better so to arrange the task and its lighting as to make the necessary detail (for instance a fault in cloth) conspicuous by means of shadow or colour contrast, rather than to increase illuminance to very high levels.

However, there are still difficulties in applying laboratory data since although watch repairers do seem to need quite high illuminances on certain parts of the task even if they are using magnifying glasses, when reading small print whose critical detail may at first analysis appear to be rather similar in size and contrast, we can often do quite well at modest illuminances. It is often said that reading is a 'familiar' task, and so it is to most of us, but watch repairing is presumably familiar to professional watch repairers as well.

In fact 'familiarity' does turn out to have some bearing on the problem, but in rather a complicated way.

UNCERTAINTY

If an observer is presented with an array of lights, any one of which may flash, and is asked to respond to each flash by pressing an appropriate key, there being one key for each possible light, it is found that reaction time is a linear function of the logarithm of the number of possible lights (Fig. 3).[7] This is called Hick's Law and is usually discussed in terms of Information Theory, in which the human subject is regarded as a communication channel between the stimulus or input and the response or output. Like all

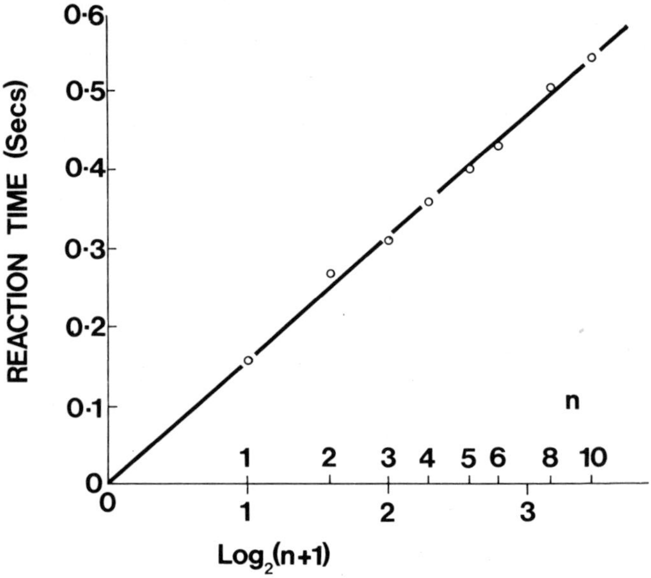

FIG. 3. Choice reaction time as a function of the number of equally probable stimuli (n).[7]

communication channels the human nervous system introduces noise or distortion into the system. Further, it has a limited channel capacity and not all the information in the input necessarily appears in the output. The information content of the stimulus may be measured by the amount of uncertainty it removes. If there are only two possible stimuli, each is said to contain only one 'bit' of information, since a single choice will define the result. If there are four possible stimuli, each contains two bits of information since two successive choices (first, it is a, b or c, d, second it is a or b etc.) will define the input. Similarly, if there are eight choices the input contains three bits of information and so on. To generalise, if the input must be drawn from a set of N equally probable stimuli its information content is $\log_2 N$ since this is the minimum number of simple choices required to define one stimulus from the set without ambiguity. When all stimuli are not equally probable the information content is reduced, but for a more formal mathematical treatment of complex cases the reader is referred to references 8 and 12.

To return to Hick's Law, if there are N lights there are $N + 1$ possible choices, the additional choice being that no light at all has flashed. In Fig. 3 it will be noted that the abscissa is scaled as $\log_2(N + 1)$ and Hick's Law is normally expressed as 'choice reaction time is a linear function of stimulus uncertainty'.

Boyce[9] has discussed some apparently anomalous results obtained by Rowlands in a Landolt ring test. Rowlands used two ring types, a simple ring (type A) and a more complex multiple ring (type B) in which the gap was hidden in a maze-like figure. It was found that although performance for the type A ring was higher than that for the type B ring, (implying that the type B task was more difficult than the type A task) performance for the type A task started to decline below about 300 lux, while that for the type B task started to decline only below about 30 lux. This is contrary to the normal expectation that performance of difficult visual tasks is more sensitive to illuminance than performance of easier ones.

In discussing this anomaly Boyce distinguishes between the 'difficulty' and the 'complexity' of visual tasks. The type A task was more difficult since the critical detail was smaller, but the type B task was more complex. As used by Boyce the idea of complexity corresponds quite closely to that of uncertainty.

It would therefore appear that visual performance is related not merely to the size and contrast of the critical detail, but also to the uncertainty of the task.[22]

DISCRIMINATION

If observers are asked to discriminate two lines of significantly different lengths presented individually, they do not make mistakes. As the number of different lengths they are asked to discriminate increases confusions appear and a limit is ultimately reached at which a further increase in the number of stimuli only increases the likelihood of confusion in the response. The results are apparently unaffected by the range from which the differing stimuli are selected, so long as the differences between each separate stimulus are well above threshold. In terms of Information Theory, the number of different stimuli which can be discriminated without confusion represents the channel capacity for that kind of judgement. It is measured, as before, in bits, and is $\log_2 N$ where N is the number of separate stimuli which can be identified without error.

Figure 4[8] is taken from an experiment on absolute judgement of the position of a pointer on an ungraduated line—that is, a judgement of linear interpolation. Four different numbers of pointer positions were used, 5, 10,

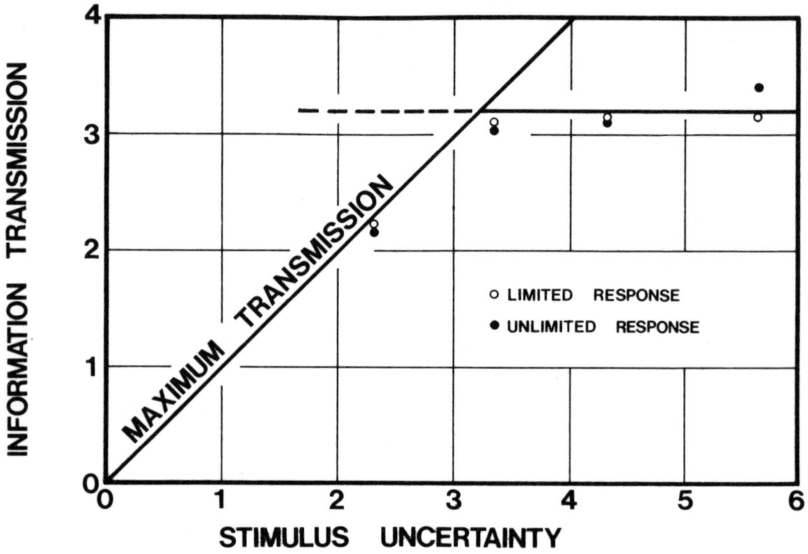

FIG. 4. Information transmission as a function of stimulus uncertainty for judgements of the position of a pointer on a line. With limited response there were as many response categories as stimuli (see text). With unlimited response there were always 101 response categories regardless of the number of stimuli.[8]

20 and 50, and two different response conditions. In one response condition the subjects were always instructed in the set of possible positions which could occur and were allowed to use no other responses. In the second condition observers were allowed to describe the position as any integer from 0 to 100 and were not instructed as to how many possible positions in fact there were. Both kinds of condition gave similar results; the channel capacity for this kind of judgement is apparently 3·2 bits, which means that only $2^{3\cdot2}$ (*i.e.* 10) positions can be discriminated under these conditions.

This may seem very surprising in view of what we normally think we can see, but it must be remembered that this experiment is on absolute judgement of linear interpolation, not the performance of a common task.

Figure 5[10] is on absolute judgements of the positions of dots in a square—that is, employing two spatial dimensions rather than one. The channel capacity is still limited, but is somewhat higher for two dimensions—at 4·4 bits (that is, $2^{4 \cdot 4}$ or 19 discriminable positions). Most work in this field has been carried out on comprehension and discrimination of sound. This is hardly surprising in view of the origins of

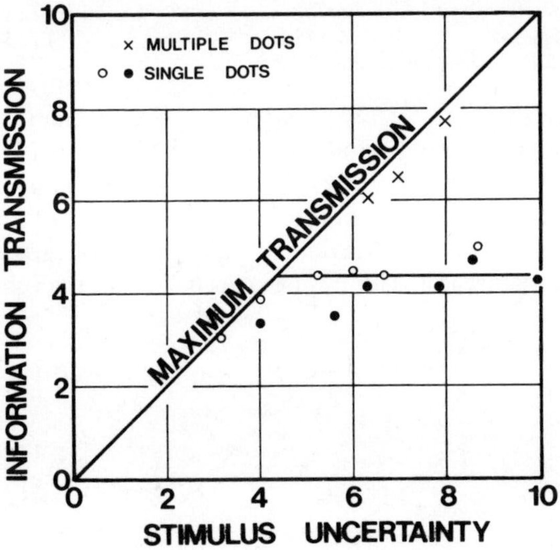

FIG. 5. Information transmission for absolute judgements of positions of dots in a square. Multiple-dot data and open circles are from Klemmer and Frick; filled-in circles are from Osborne, Quastler and Tweddell.[10]

concepts such as information and channel capacity, but Table 1 (11) shows results for visual stimuli which differed in size only, hue only or brightness only, and then for any two of these variables, and finally in all three. In all cases, adding a second or third dimension increases the channel capacity, but not by a process of simple addition of the channel capacities taken individually. It has been suggested that, for absolute judgements at least, maximum information transmission will be obtained when no more than three alternatives are used with a single variable, but with many variables involved.

TABLE 1
INFORMATION TRANSMISSION FOR VISUAL STIMULI VARYING IN ONE, TWO OR THREE DIMENSIONS

Stimulus dimensions	*Information transmission (bits)*	*Equivalent no. of absolutely discriminable stimuli*
Size	2·84	7·19
Hue	3·08	8·45
Brightness	2·34	5·06
Size–hue	3·55	11·90
Size–brightness	2.98	7·89
Hue–brightness	3·76	13·55
Size–hue–brightness	4·11	17·28

Returning to Fig. 5, the crosses are data from experiments in which more than one dot was presented simultaneously (up to four in a 3 × 3 grid) and each possible separate pattern was treated as a separately discriminable stimulus. Simultaneous presentation is another way of increasing the number of dimensions of judgement, and leads to greatly increased information transmission. This implies parallel rather than sequential data processing.

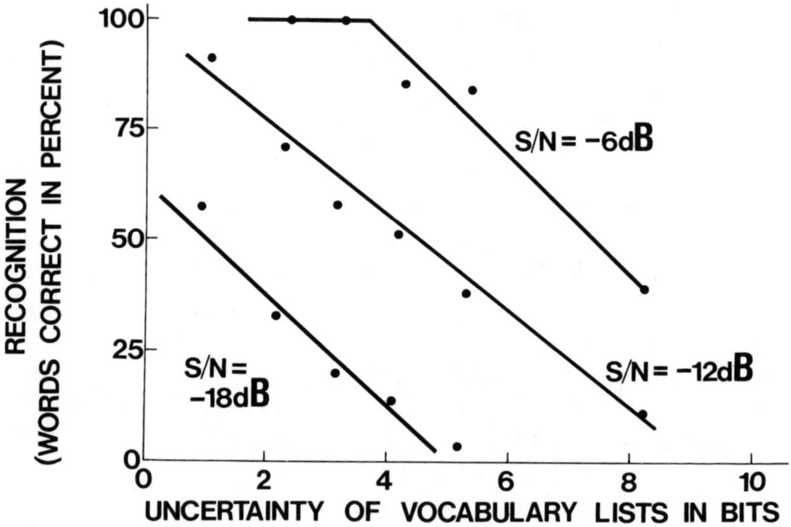

FIG. 6. Recognition of auditory monosyllables presented against different values of signal-to-noise ratio. For each value of S/N the curves show performance as a function of the number of words in the test vocabulary.[13]

Experiments of this kind are sometimes rather hard to interpret, and for all but the simplest cases the mathematics rapidly become very daunting. Nevertheless they should at least make us cautious in supposing that classical concepts such as 'critical detail', taken by themselves, are likely to be of much help in predicting performance in real situations.

RECOGNITION

It is known that at least in auditory recognition, performance is a function of the number of alternatives, and of the signal-to-noise ratio. Figure 6[13] is the results of an experiment on recognition of spoken words with different levels of masking noise. The test vocabularies consisted of sets of 2, 4, 8, 16, 32 or 256 monosyllables, corresponding to uncertainties ranging from one to eight bits. One word was read at a time and the subjects always knew from which list the words were being selected. Figure 7[14] is from an

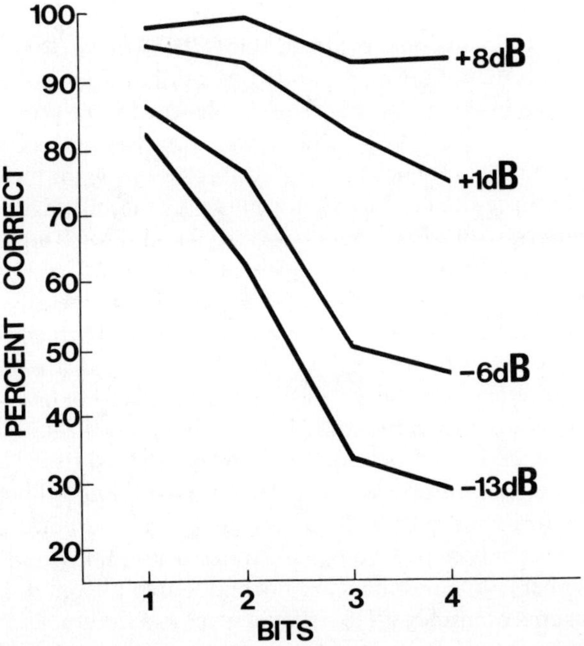

FIG. 7. Recognition of nonsense disyllables presented against different values of signal-to-noise ratio. For each value of S/N the curves show performance as a function of the average rate of input information in bits per item.[14]

experiment on recognition of nonsense disyllables constructed in such a way as to obtain signals of known uncertainty. The author knows of no corresponding work in visual perception, but the similarities of Figs. 6 and 7 to curves of visual performance under reducing contrast are apparently striking, and research on these lines in vision would be well worthwhile.

Cole[23] has suggested that the concepts of Information Theory may be relevant to street lighting.

THE IMPLICATIONS OF INFORMATION THEORY

The very brief description of work on Information Theory given above points to certain inadequacies in the approach to visual performance hitherto adopted by workers whose primary aim has been to obtain results which might be relevant to lighting practice. If similar concepts were to be carefully applied to the very confusing results we do have on visual performance it is possible that something of genuine practical value might emerge.

Some general conclusions are already clear. It is known, for instance, that in reading adults do not normally study the details of each individual letter, but instead take in the gross shapes of whole words and groups of words. The difficulty of accurate proof reading also indicates that recognition of a word will be highly influenced by the context. This again implies that we cannot hope to predict reading performance by attempting to equate the details of letters with visual detail used in a Landolt Ring experiment. Furthermore, reading is a multi-dimensional task with rather low uncertainty, since there are only a few letters with which any given letter might be confused and probably even fewer words with which a given word may be confused. From Figs. 6 and 7 we should therefore expect reading performance to be fairly resistant to conditions of low contrast and bad lighting, as we find in practice. In checking lists of random numbers, however, the uncertainty for each number may be relatively low, but the pattern of groups of numbers gives us no help; we should therefore expect performance to be more critically affected by contrast and lighting.

However, it is only in psychological laboratories that people check pure random numbers—the numbers are normally drawn from a more limited set, such as sums of money all less than a given maximum, and often more than a given minimum. Performance should then be less affected by lighting and contrast than in the pure random case.

Information Theory provides us with new criteria with which to classify visual tasks and suggests new kinds of experiment. Further, as will be shown later, this approach to discrimination and recognition may also be of help when we wish to use lighting not to show what is really there as plainly as possible, but deliberately to manipulate perception in a controlled way.

COLOUR DISCRIMINATION

It is generally realised that colour vision enables us to discriminate between many different surfaces and objects whose overall reflectance is the same, and which appear similar in a black and white photograph. It therefore adds several additional dimensions of judgement to visual discrimination (Table 1), especially when we take into account not merely hue and chroma of homogeneous surfaces, but pattern and texture which give rise to a fine structure of colour differences in a given surface. Information Theory adds little to our understanding of these phenomena, except to indicate the importance of having as many different dimensions of judgement as possible since our channel capacity for each single dimension turns out to be so surprisingly small.

However, the concept of 'visual clarity' may be reconsidered in these terms. It has been found that sources which give a large separation between colours on the CIE uniform chromaticity chart can give equal satisfaction at lower illuminances than sources which give a smaller separation on the CIE chart.[15] Compare this kind of result with Fig. 2, from which it appears that the perceptual difference between surfaces of different reflectance increases with illuminance. If we think of perception as an active process by which we try to make sense of and deal with our surroundings, it must be advantageous if the various possible dimensions of judgement are as clearly separated as possible.

The idea of maximising dimensions of judgement as an aid to perception should make us think more carefully about the kind of colour rendering we require from our light sources, since, as has been pointed out by Lynes, the arbitrary equation between 'good' colour rendering and similarity of spectral distribution to some phase of daylight or sunlight may be inadequate. In particular, the effects of 'visual clarity' still appear when sources of 'bad' colour rendering are employed, so long as they maximise the colour gamut. However, it is unlikely that we can go very far in this direction since people do not like familiar coloured objects to be too distorted in appearance, especially food and the human features whose colour appearance is so often taken as an indication of health and well-being.

COLOUR CONSTANCY

Lighting engineers normally learn all about CIE standard sources, colour mixture equations and so forth, and grow up with the knowledge that the spectral distribution of the light reaching our eyes from any coloured object is drastically affected not merely by the spectral reflectance curve of its surface but also by the spectrum of the source. Despite this, gross changes of spectral distribution do not often lead to great distortion in the perceived colour of different surfaces, which has led to great difficulty in producing meaningful indices of colour rendering. We know that many sources which are perfectly satisfactory for normal use have rather poor indices both on the CIE system and even on the Crawford System, which at least seeks to take account of colour difference thresholds.

Just as with lightness constancy, it can be shown that the known processes of adaption, reaction and inhibition in the retina resulting from coloured stimuli can be used to account for the relative stability of our perception of colour, despite gross changes in the incident light. Theories of this kind are normally described as 'opponent colour' theories, and are often dismissed by people whose primary interest is in physics as being inconsistent with the trireceptor theory. However, as explained in considering neural interaction in the retina as a possible explanation of lightness constancy, there is no reason to suppose that the type of results we obtain in laboratory experiments using small unstructured fields will necessarily be the same as in normal perception when the eyes scan a complete visual field with differences of contour, lightness and colour. The elementary phenomenon of simultaneous contrast indicates that some opponent processes must exist.

CONDITIONS FOR ACCURATE PERCEPTION—2

In our first discussion of this problem we introduced two rather vague statements—the first condition merely required that the illuminance should everywhere be high enough to perceive 'all necessary detail'. The second that the spectral composition of the light shall be such as to permit 'appropriate colour discrimination'. Has the resulting discussion of the concept of 'critical detail' in relation to information theory and of 'visual clarity' helped to make these statements more precise? Not necessarily, it may even have created confusion by casting doubt on whether we know what information we need from the environment, and destroying some of our familiar certainties.

Nevertheless, lighting is an essentially practical subject in which theory lags years behind practice, while the same is true of architecture and interior design. We should be fairly confident that current practice in lighting has, by trial and error, come into accord with current practice for interior design and that illuminances within the customary range of 200–400 lux are likely to be sufficient for accurate perception of the elements of an interior. The most common fault in many bread-and-butter lighting schemes is an underlit ceiling, resulting from the use of fully recessed luminaires and relatively dark floor cavities, but even in such circumstances we rarely fail to perceive all relevant details of the ceiling, at least in working environments. The problem is that the balance of brightness throughout the interior seems unsatisfactory. Quite suddenly we seem to have stopped talking about the psycho-physics and psychology of perception, but have come back to the customary jargon and preoccupations of lighting engineering, but is this surprising? This discussion started by pointing out the contrast between the conclusions we might draw from our common experience of lighting, and the conclusions which some people seek to draw from certain highly artificial experiments under controlled conditions. It should have been clear all along that our common experience is much more soundly based than controlled experiments, which doubtless give sound information about *something* but not necessarily what lighting engineers need to know.

Current practice in lighting, as represented, for instance, by the recommendations of the IES Code,[16] lays stress on the need for an adequate quantity of light, appropriate choice of reflectance of room surfaces, directional effects, suitable colour rendering and the avoidance of disagreeable side effects such as glare and flicker. Figure 8 is reproduced from the IES Code and shows the recommended range of reflectances and illuminance ratios as between walls and ceiling on the one hand and working plane on the other. The minimum horizontal illuminance recommended for any place where people work is 200 lux, and this figure, taken in conjunction with Fig. 8, will ensure that at least conditions (a) and (b) on page 86 are satisfied. Colour rendering is not specified in quite the same way, but again, the Schedule in the Code lists those types of sources which have been found acceptable in practice for various situations, and this comes to the same thing.

All the normal recommendations of lighting engineering, adequate uniformity, relatively high reflectances and so forth will tend to reduce the variations of illuminance in an interior and so promote speedy and accurate perception. We appear to have come full circle, since by investigating the visual process in a highly sophisticated way we have apparently learned

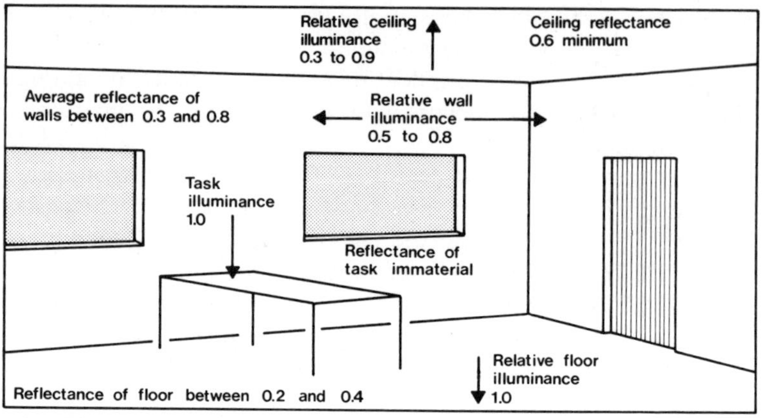

FIG. 8. The recommended range of room surface reflectances and relative illuminance ratios for normal working environments recommended in the IES Code.[16] (Courtesy The Lighting Division of the Chartered Institution of Building Services.)

nothing which every lighting engineer does not already know. However, on the way we have abandoned a number of things which many lighting engineers believe but hardly ever use in practice, such as the importance of critical detail, the desirability of high illuminance even for quite common tasks, and the belief that brightness is the perceptual correlate of luminance.[17]

However, there is another aspect of current practice in lighting which is hardly discussed in the Code—display, the theatre, lighting for dramatic effect. Everyone who has ever been to a discotheque knows that in such a place current practice is quite different. The surroundings are generally rather dark, except for the stage or the dance floor. The maximum use is made of spotlights, often in strong colours, 'strobe' effects and so on. Even in normal display situations or expensive restaurants where trick 'effects' are not employed there is the same tendency to use dark surroundings and to confine virtually all the light to a relatively small number of objects—the tables in a restaurant, the goods for sale in a shop. It is true that in such places systematic methods of lighting design are rarely used, and those who determine the lighting may not be lighting engineers as we normally understand them. Nevertheless the consistency with which all the normal recommendations of good practice for working environments are stood on their heads should surely make us think.

It is often said, of course, that lighting for the theatre and display is an 'art', as opposed to the more mundane engineering applications, but this does not explain anything at all. It is true that the objectives of design are different, and some of the people who practise this kind of lighting may be described as artists as opposed to engineers, but in what does the art consist, and why should it so often involve the exact opposite of engineering? Perhaps we should use the term 'artifice' rather than 'art' since this more accurately describes what display lighting technique is designed to achieve. It is not, as in working environments, simply to show us what is there, but to manipulate our perceptions in a controlled way. Some things should be made to seem especially interesting and exciting, others are scarcely intended to be seen at all. It is usually obvious which class of objects should fall into which category. For selling it is the objects for sale which must be emphasised, in restaurants the tables and perhaps the bar, in the theatre the actors and settings, in discotheques the dance floor and other special displays. In none of these interiors do we particularly want to see anything else except when we are moving in or out, and only enough light need be cast at the surroundings to prevent our actually bumping into things. What would happen if we lit a working environment like that? Would it then seem equally exciting? Again, common experience indicates that it would not; the use of downlighters in certain types of interior is now very common, but they are not by any means successful. We say that the room appears gloomy, and recommend that the walls and ceiling shall be adequately lighted. There is obviously something more to it than has yet appeared. Although strong contrasts of illuminance are commonly used to assist in building up a dramatic effect, such contrasts do not, by themselves, assure a successful result. One cannot but be reminded of the situation in lighting engineering about 15 years ago. There were rumours from the US of 1000 and 2000 lux installations in offices and how good they were. Then people started occasionally to employ such illuminances in Europe and sometimes the effect was striking and sometimes a failure. It turned out that high illuminances were sometimes attractive, although always expensive, but for success a great many other conditions had to be fulfilled as well.

It may be that, in display lighting, although high contrasts are a *necessary* condition for a dramatic effect they are not a *sufficient* condition. Can study of the perceptual process help us to pursue the problem?

PERCEPTION OF OBJECTS

The visual system appears to be organised in such a way as to be especially sensitive to contours and sharp changes of colour or lightness in the environment. This goes a long way to explain both *lightness constancy* and *colour constancy*, terms which describe the fact that our perception of the nature of a surface remains remarkably stable despite large changes in the stimuli. There are, however, many other perceptual constancies which are not so easy to explain in the present state of knowledge. These include size constancy and shape constancy; as we approach or recede from an object its size does not appear to change although the size of the retinal image must change in accordance with the Law of the Visual Angle. Similarly the shape of the retinal image of any object other than a sphere must change as its orientation changes, and yet objects generally appear to be of constant shape. It is difficult to conceive of any kind of retinal organisation which could account for these constancies and in fact, research into the behaviour of young babies appears to indicate that not until the second or third year does the child acquire concepts of solid objects of permanent size and shape persisting in space and time.[18]

The way in which these concepts develop and to what extent certain kinds of response are due to neural organisations present at birth is still largely obscure. There is no difficulty in believing that lightness and colour constancy, to the extent that they are the result of retinal organisation, are present at or very shortly after birth. The problem arises with those aspects of perception in which the concept of meaning arises. Figure 9 shows the well-known Necker Cube, whose perceived orientation changes spontaneously from time to time as we look at it.

In Fig. 10 the three rectangles are exactly the same size, but they appear to be of different sizes since the drawing implies that they are at different distances. Most people are familiar with photographs of Ames' rooms, built in falsified perspective so that a child can look larger than an adult because we are misled as to the real distance of each. Nevertheless, we are assured that the Ames' illusions only hold where the people seen in the room are not well-known to the observer; wives are said never to be deceived as to the true stature of their husbands!

It is possible to conceive that at least some of whatever neural mechanisms underly the Necker Cube illusion may be present at birth, but not the mechanism of the illusion of Fig. 10. Research information is voluminous but we are still only able to say, rather weakly, and not unlike the 17th century empirical philosophers, that human concepts of permanent and stable objects arise in infancy as a result of interaction between the baby and its environment, in which information obtained from all senses is progressively assimilated and generalised. In particular, vision by itself can give us no concept of solidity, which must come from touch,

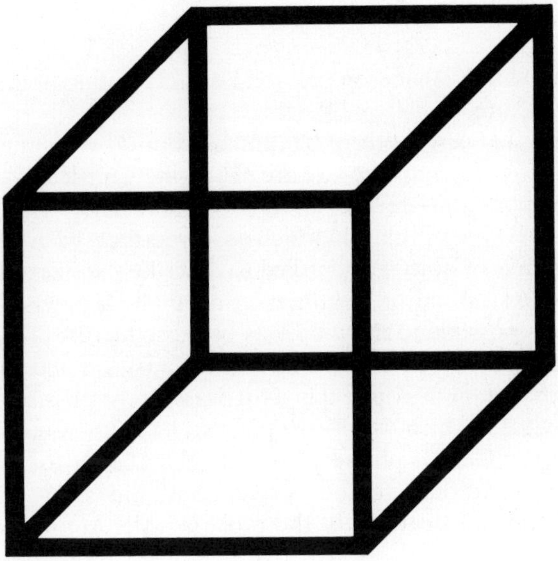

FIG. 9. The Necker Cube. The orientation changes spontaneously as we stare at it.

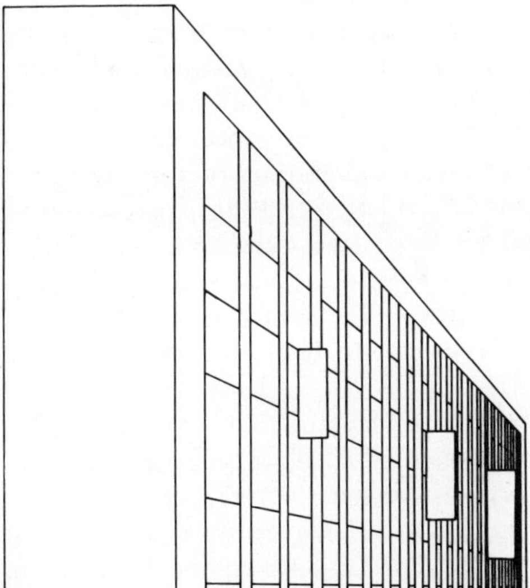

FIG. 10. The perceived size of the three identical white rectangles is affected by their background.

and yet adults will say that they 'see' solid objects and do not need to touch them to make sure.

It is obvious that perception of size and perception of distance are closely inter-related (Fig. 10) and when we are deliberately misled as to the one we will be correspondingly misled as to the other. There is a voluminous literature on this kind of thing, in which observers have been presented with misleading arrays of objects and asked to describe the size and distance of each. In any given situation the observer normally sees what makes most sense, and it is possible to draw up lists of those features of a visual field which are used as 'cues' or 'clues' for relative position and therefore size. Such research is now less popular as it was rarely possible to find any particular feature which is over-riding in its perceptual significance. Nevertheless the Gestalt philosophers, in their search for simple neural organisations to account for our perceptions, did draw attention to a number of important features of the process which we would do well to remember.

In particular, perceptions seem to be of 'wholes' and not parts in the sense that we do not perceive figures as being made up from simpler elements. The figure 4 is perceived as a figure on its own and not as figure 1 or + with additions. After the Gestalt school it was very difficult to go on believing that complex 'perceptions' are really built up from elementary 'sensations', which is a great relief. We can now conceive of an elaborate information processing system which takes in data of various kinds and processes it along parallel paths very rapidly, resulting in perceptions such as 'red chair'. It is also clear that the visual data in the optic nerve cannot be described in terms of elementary physics as little dots of luminance, but is concerned with significant features of the scene, such as contours, angles, colour changes and gradients etc. In later discussion we shall refer to these elements as the 'visual signs'.

It is also important to understand that the kind of maturational learning which leads to concepts of solid objects is apparently non-reversible and results in neural organisations in the visual system which have some of the characteristics of a logic circuit. It is necessary to lay some stress on this since there are those who will assert that we do not, for instance, 'see' a table, but only a brown parallelogram which we 'think' is a table. The process of perception is not like this at all; perceiving a table is the primary experience—we are only induced to say that what we actually 'see' is a brown parallelogram after a process of analysis. It is absurd to suggest that the results of the analytical procedure are in some way more 'real' than the primary experience itself.

THE FIGURE–GROUND PHENOMENON

Figure 11 is very well known. It can be seen either as a black vase on a white ground or as two white faces looking at each other on a black ground. We can, by effort of will, hold our perception of either one or the other but in casual vision the two may alternate. We do not have both perceptions simultaneously, although they may at times alternate very rapidly. The point about Fig. 11 is that the contour round the black shape in the middle is always seen *either* as the outline of a vase *or* as the outline of a face but

FIG. 11. The figure–ground phenomenon. Does this illustration represent two faces on a black background, or a black vase on a white background?

never both at once, although it is apparently possible to see one side of the contour as the outline of a face and the other as the outline of a vase. Nevertheless, a contour is perceived as having one side only, that of the object which it bounds. The background is perceived as extending behind the object. This is called the 'figure–ground' phenomenon, and describes our tendency to perceive anything within a closed contour at which our attention is directed as an object in itself, and any stimuli surrounding this contour are perceived less clearly and do not have this object quality. We can, of course, direct our attention successively to a particular object and then to its background, and in real life objects and backgrounds are often at different distances so that they cannot both be in focus together, although no such problems arise with Fig. 11.

This is, of course, quite consistent with what we know of eye movements and the fovea of the retina, and in a loose sort of way it supports the concept of a scanning mechanism in perception, although the analogy between the eye and the television camera will be found very misleading if we attempt to use it inappropriately.

EXPECTATION

Our perceptions are influenced by what we expect to perceive and sometimes we persist in perceiving what we expect in defiance of the stimulus. At an anecdotal level, husbands who fail to notice that their wives have changed their hair style have for long been a standing joke, and this kind of thing has been investigated by psychologists. One of the best known experiments consisted in presenting playing cards to the subjects for controlled short periods of up to one second and asking them to identify them.[20] Some of the cards were normal but for others colour and suit were reversed (i.e. black hearts and red spades). It was found that the minimum time required to identify the incongruous cards correctly was greatly in excess of that required for the normal cards. In this experiment all cards were represented for longer and longer periods until correct identification occurred, but in real life we do not normally do this; we glance at something and, unless it seems important to us, look at something else. We can often be deceived (Fig. 12).

Some of the phenomena studied by Gestalt psychologists can be discussed in the same terms. For instance, at a quick glance a broken outline may appear to be complete, especially if its approximates to a circle or other

FIG. 12.

regular figure. Similarly, an irregular shape (for instance, a hexagon) is likely to be recalled as a regular one, unless there has been reason to study it very carefully. These and many other phenomena originally presented as the Gestalt 'Laws of Organisation'[21] can be discussed in terms of Information Theory. If presented with ambiguous or unusual stimuli the observer is likely to perceive something which requires *less information* to specify it than the stimulus itself. A regular polygon can be specified fairly easily; an irregular one has sides of at least two different lengths. To specify a broken circle we must first describe the circle itself, and then the position and size of the break—and so on. As we have seen, our capacity for processing information is limited. It is entirely to be expected that under conditions of casual or otherwise sub-optimal viewing the limited amount of information which we can process in a given time should be accepted as the complete perception.

EXPECTATION VS. SURPRISE

What do we mean by saying that we are 'surprised' to see something? This must be that we have compared what we do see with what we expected to see and are aware of some mismatch. However, the phrase 'what we expected to see' implies a very great deal, and in considering it further it is necessary to be very cautious in our use of language. The meaning is fairly clear when we are revisiting some place which we know well—our own living room, for instance. We already know what this is like, although our description of it would not necessarily be accepted as accurate by another observer. Psychologists generally use the term 'schema' to describe whatever set of mental concepts results from our 'knowing' our own sitting room. We would only be surprised on entering this room if we perceived something entirely inconsistent with the schema—the furniture had been moved or taken away, and so on. However, we might also say that we were surprised on entering some other room which we had never seen before, and this is more complex. For instance, the room in question might be in a house we had never previously entered, although we knew it was occupied by somebody we had always thought of as rich. If the room were poorly furnished we would regard this as sufficient reason for surprise. In other words, our schema of a rich man implies a great deal more than thinking he has a large bank account, but includes all sorts of expectations about his house, other possessions and life style. In fact, all our schemata appear to generate a wide range of expectations as to future perception which is extremely important for design and underlies much of 20th century architecture, although it is sometimes expressed in a tortuous and incomprehensible jargon.

Most theorists would say that the schemata generated by a building should be consistent and helpful to the occupants. An office block should not look like a Renaissance palace, while an airport terminal building should 'evoke the idea of flight', and so on. This is all very well for those whose schemata of buildings are highly developed, but does not always work for those who have never thought much about the theory of design, and have seen so many office blocks got up to look like Renaissance palaces that their schema of an office block does in fact include a palatial facade. This kind of thing should always be borne in mind when discussing appearances. For instance, it is no good telling somebody that fluorescent tubes are the best way of lighting a domestic kitchen if their schema of fluorescent tubes is firmly linked to office blocks and shops and they do not like to think of their kitchen as a workplace. If the mismatch between their schema of a fluorescent tube and that of their own kitchen is very resistant to change they will always be unhappy with the tube. However, schemata sometimes do change with time, (usually called a change in 'fashion') and it generally pays to be neither too far ahead, nor behind.

PERCEPTION UNDER SUB-OPTIMAL CONDITIONS

An imposing entrance hall in a building implies equally imposing corridors leading away from it to large and well furnished offices. We are likely to be satisfied if such expectations are fulfilled but may feel some unease if they are confounded. Under sub-optimal conditions of viewing, when the lighting is bad, we continue to generate these expectations of future perception, but in this case we may not receive sufficient visual data continually to cross check our expectations. In such a case our schemata, or expectations, come to have a greater influence on our perceptions themselves.

Information Theory can help us to understand how it is that perception appears to remain fairly stable over a wide range but when the conditions of lighting fall below some critical level our ability to perceive accurately falls off rather rapidly, resulting in a 'knee' in the visual performance curve. As explained earlier, it is known that discrimination is most accurate between stimuli which differ in a large number of dimensions, but as illumination falls, not merely does the differentiation between the various dimensions reduce, some of them vanish altogether. For instance, in scotopic vision we can distinguish only brightness but not colour, and even under mesopic conditions the range of our colour responses is reduced and we cannot employ the additional dimensions of texture since we are unable to resolve the small differences in colour and brightness which contribute to them. Again, as acuity reduces we may still be able to resolve major contours, but not to distinguish critical signs, such as the distinction between rounded and sharp corners. For each kind of discrimination we should expect to find some form of threshold. When the viewing conditions are good enough to bring all dimensions into play, visual performance would then show only minor improvements with further increase in illuminance.

PERCEPTION UNDER HIGH CONTRAST

Consider a room in which a small statue is illuminated very brightly by means of a profile spotlight giving no spill light, with no illumination elsewhere. We can see all the details of the statue very clearly; the illuminance is high so our acuity will itself be high, vision is photopic so that we have all the dimensions of colour available. We now stand back from the statue and look around the room.

It can be shown[6] that if all direct light is confined to one object (the target) of area A_t, if the total area of the background (the bounding surfaces of the room) is A_b and its average reflectance ρ_b, then contrast between target and background defined in the usual way as

$$(L_t - L_b)/L_b = \frac{A_b(1 - \rho_b)}{A_t \cdot \rho_b}$$

In the case we have described the statue might have an area of about 1 m^2 and if the room is a 3 m cube A_b/A_t will be 54. Referring back to Fig. 2, where illuminances differ by a factor of no more than 10 or so the Apparent Munsell Value in the brightest part of the room is little different from that in the darkest part; that is to say, lightness constancy will hold. Suppose that both the statue and the room in which it is placed are white, $\rho_b = 0{\cdot}7$. Luminance contrast and illuminance contrast will then be much the same, and will come to about 23 and again Fig. 2 indicates that our perception of both the room and the statue are likely to be veridical.

If, however, the room is dark in colour, with an average reflectance of about 0·1, the luminance contrast will be nearly 500. It is difficult to apply Fig. 2 to a case differing so greatly from the conditions under which the curves were obtained, but it is clear that our perception of the room will be greatly impaired. The luminance of the bounding surfaces will be low, such contrasts as there are between surfaces will be hard to detect, while it will be difficult to resolve contours. Apart from this, phototropism will make us continually glance back at the statue, so that as we look about us we find it difficult to adapt to the background.

If the statue is part of an exhibition which we are visiting we may well say that the lighting has made a splendid contribution to the display. Whatever processes lie behind the figure–ground phenomenon tend to heighten our appreciation of the statue; it is, after all, what we want to see, we are not much interested in the room and the fact that we can hardly see that does not worry us at all. We will pass on wishing that the lighting in our office was as exciting as this. Referring again to Fig. 2, we note that at high illuminances light surfaces take on some of the qualities of a self-luminous surface; we may tell our friends that the lighting was so clever the statue seemed almost to glow. Suppose, however, that the statue was taken from the exterior of a building which we know well. We may be aware that the sculptor intended it to be seen by daylight against a background of colour and texture similar to itself and surrounded by other statues. Purists are often irritated by wrenching an individual object out of context like this; we may say that we wish those designers would not try and be so clever with their display techniques and that a lot of detail has been lost because the shadows are so dense.

Suppose again that we are not a visitor to the exhibition at all, but have called to see whether the carpet is properly laid. This carefully arranged 'dramatic' lighting on the statue is then merely an irritation.

The explanation of these various possible responses to the same scene should now be clear. Under conditions of high contrast all the usual processes which aid veridical perception function to make the brightest object seem 'special', 'exciting' and so forth. It is seen to be what it is, but at the same time something more. Light or strongly coloured surfaces stand out especially well, and our tendency to structure each part of the visual field into figure and ground is reinforced where we can only see one particular 'figure' and are quite uninterested in the 'ground'. At the same time our perception of the background is impaired. We cannot easily resolve colours, contours and texture and are therefore deprived of many of the normal visual signs. Even such signs as we do receive are not distinct or clearly differentiated from one another. Our ability to recognise and discriminate between different objects is therefore greatly impaired, much more so than might be indicated by studying curves of visual acuity or colour discrimination.

In such circumstances our perceptions become much more influenced by suggestions and expectation, which is the basis of much design. A basement room with plain brick walls, painted in dark colours, if lighted with a few spotlights, becomes a nightclub, an exciting place in which we will be glad to pay exorbitant prices for refreshment and submit to acute discomfort under the impression that we are dancing. Wooden frames covered in painted hessian, can, if placed on a stage and viewed appropriately from a distance, be a gorgeous palace. The same stage set under working lights and seen from close to can be very disappointing.

This kind of thing is often called the 'art' of lighting. As described here it may sound merely like the explotation of a few simple tricks but might be worth dignifying by the title 'art', or at any rate, 'good design' if it serves a clear and appropriate purpose. Engineers often talk and write as if 'art' is dispensed by artists rather as medicines are dispensed by pharmacists and, in their different way, artists themselves sometimes make the same mistake. The discussion of the lighting of the statue given earlier indicates that however clever or 'dramatic' the lighting, the result will not be regarded as successful if the viewer does not recognise that the total design is appropriate to *his* purpose. Of course we all have many different purposes at different times and often at the same time, so that we can sometimes make a compound judgement and say both that the display technique was very clever but it irritated us because we could not see clearly.

A corresponding error is often made in attempting to 'appraise' some aspect of interior design. If taken into rooms, or shown pictures of them, observers may commit themselves to all kinds of opinions about their appearance which can be submitted to analysis of one kind or another. However there is all the difference in the world between looking at a room in detachment and living or working in that room, and we should be careful that, in the name of science, we do not reinforce the architectural tendency to regard spaces which should have a real function as formal pictures which can be assessed from the outside.

Perception has been described as 'the search for meaning' and it is obvious that the meaning of a space varies enormously with the purpose of the occupant. Display lighting technique works best for the viewer who is primarily a tourist; a visitor to the theatre, a restaurant, an exhibition. In these cases his aim is primarily to be excited or entertained. We should never confuse this attitude with the very different aims of people who wish to live and work comfortably, although we should enable them to do this as pleasantly as possible.

CONCLUSIONS

It is commonplace that the unity which is thought to have been a feature of people's view of the world in the 18th century has been lost as a result of the great changes and discoveries of the 19th century. New studies and branches of studies arise every few months but increase in knowledge does seem to have increased our understanding. In this chapter I have sought to show that we can now at least begin to relate together certain studies which have hitherto been regarded as largely unrelated. At the simplest level 'functional' lighting can be related to 'dramatic' lighting once we understand that the former is assisting those features of the perceptual system which promote veridical perception, such as lightness and colour constancy, while the latter is exploiting other features, but in such a way as to manipulate perception for the designer's purposes. Whether or not the result will be regarded as successful must depend on whether the designer has correctly interpreted the viewer's needs.

Again, psycho-physics appears to give us one kind of information about the visual process, psychology another, and common experience another still. Until quite recently it was difficult to relate any of these kinds of information to each other, but we are now beginning to see how this might be done.

Moreover, engineers are accustomed to distinguish themselves sharply from artists, to such an extent that meaningful communication becomes very difficult. We should now be able to see that engineers have as their aim the satisfaction of one kind of human purpose and artists another, but related, kind. The lack of communication is due at least partly to the engineer's tendency to attempt to exclude human purpose from his objectives which he redefines in terms of physics, but in this he often fails. Just as the artist must learn that the laws of physics will not bend themselves to his will, so the engineer must learn that perceptual systems are not physical measuring instruments. When both have learned such lessons well enough communication may again be possible.

REFERENCES

1. GIBSON, J. J., *The Senses Considered as Perceptual Systems*, 1968, Allen and Unwin, London.
2. LOWRY, E. M. and DA PALMA, J. J., *J. Opt. Soc. Amer.*, 1961, **51,** 740.

3. Hubel, D. H. and Weisel, T. N., *J. Physiol.*, 1962, **160** 106. (Reported by Dodwell, P. C. in *New Horizons in Psychology*, (ed. B. M. Foss), 1966, Penguin Books, London.)
4. Saunders, J. E., Ph.D. Thesis, University of London, 1969.
5. Hurvich, L. M. and Jameson, D., *The Perception of Brightness and Darkness*, 1966, Allyn and Bacon, Boston.
6. Jay, P. A., *Ltg Res. Technol.*, 1971, **3,** 133.
7. Hick, W. E., *Quart. J. Exp. Psychol.*, 1952, **4** 11. (Reported in reference 12.)
8. Garner, W. R., *Uncertainty and Structure as Psychological Concepts*, 1966, Robert E. Krieger, New York.
9. Boyce, P. R., *Ltg Res. Technol.*, 1974, **6,** 222.
10. Klemmer, E. T. and Frick, F. C., *J. Exp. Psychol.*, 1953, **45,** 15 and Osborne, J. W., Quastler, H. and Tweddell, K. S., 'Control Systems Lab.', 1955, University of Illinois No. R-78. (As adapted in Reference 8.)
11. Eriksen, C. W. and Hake, H. W., *J. Exp. Psychol.*, 1955, **50** 153. (Reported in reference 12.)
12. Corso, J. F., *The Experimental Psychology of Sensory Behaviour*, 1967, Holt, Rinehart and Winston, New York.
13. Miller, G. A., Heise, G. A. and Lichten, W., *J. Exp. Psychol.*, 1951, **41** 329. (Reported in reference 12.)
14. Miller, I., *J. Exp. Psychol.*, 1957, **53** 388. (Reported in reference 12.)
15. Boyce, P. R., *Ltg Res. Technol.*, 1977, **9** 11.
16. Illuminating Engineering Society, London, 'IES Code for Interior Lighting' 1977.
17. Bodman, H. W., *Trans. Illum. Eng. Soc.* (*London*), 1967, **32** 22.
18. Piaget, J., *The Origin of Intelligence in the Child*, 1953, Routledge and Kegan Paul, London.
19. Gibson, J. J., *The Perception of the Visual World*, 1950, Houghton Mifflin, Boston.
20. Bruner, J. S. and Postman, L., *J. Pers.*, 1949, **18** 208. (Reproduced in *Readings in Perception*, (ed. Beardslee and Wertheimer), 1958, van Nostrand, Princeton.)
21. Hochberg, J. E., *Perception*, 1964, Prentice-Hall, New Jersey.
22. Hilgendorf, L., *Ergonomics*, 1966, **9** Part I, 31.
23. Cole, B. L., *IES Lighting Review*, 1971, **33** 42.

Chapter 5

DESIGNED APPEARANCE LIGHTING

J. M. WALDRAM
B.Sc. (Eng.), Hon.F.C.I.B.S.
Consulting Lighting Engineer, Cambridge, UK

SUMMARY

Designed Appearance Lighting is a design procedure for lighting an interior whereby a preconceived appearance of the interior can be realised. It is applicable to important interiors of fixed and stable use. The most important design factor is the preconceived appearance, and guidance on its conception is given. Its significant features, and the means whereby they are specified, calculated and achieved, are discussed in detail; important practical aspects of design, installation and maintenance are indicated. As an example, the lighting of Ely Cathedral by this procedure is described and illustrated.

INTRODUCTION

Designed Appearance Lighting is a design procedure whereby a lighting installation can be designed which will, in collaboration with the architecture, the decorations and the furniture, provide a preconceived appearance to the interior. Though it can be fully applied to only a small class of interiors, they are the most important, and call for such an approach. The procedure has advantages in indicating a design which can be relied on to produce the desired effect without extensive experiment—usually inconclusive and expensive and often impracticable. It also indicates sites for lighting equipment and probable circuit loadings at an early stage, thus permitting electrical installation, feasibility studies and cost estimates to proceed with some confidence in parallel with the more detailed design of the lighting.

CONCEPTS AND DEFINITIONS

Designed Appearance

An appearance of an interior conceived in advance which is significant and appropriate for the interior.

Appearance of an Interior

There is, visually, no architecture in the dark. The appearance of an interior results from the interaction of the architectural forms, their finishes and decoration, and the furniture with light—not, be it noted with lighting equipment, but with light as such.

Purpose

Design implies purpose. The designed appearance is meaningful only to the extent that it matches the purpose of the interior, which is possible only if that purpose is known, is specific, and is stable. Sophisticated design of appearance is futile if the use of the interior is not known fully, or is vague, or liable to change. Designed Appearance is therefore called for in only a small fraction of lit spaces: those which are architecturally significant, dedicated to clearly defined use and likely to remain so throughout their life; and where the forms, finishes and furniture are unlikely to change. Examples are auditoria, churches, some civic buildings, art galleries and museums, etc. Where the use of the interior is vague, as in hotel lounges, or changeable, as in offices, Designed Appearance is applicable only to a limited extent; though consideration of its principles is beneficial, for inept lighting may mar the appearance of an interior even when its purpose is not clearly defined.

Lighting for Designed Appearance

A scheme of lighting for an interior designed so to collaborate with its form, finishes and furniture that a designed appearance will result. It is not restricted to, nor identified with, any particular technical system of lighting.

DESIGNING THE APPEARANCE

The key to a successful Designed Appearance scheme is the concept of the appearance to be achieved, not the technical procedure whereby that appearance is realised. The technical means contribute nothing to the aims of the design; they are as effective in realising an inept scheme as an inspired one. The concept of the appearance is aesthetic, save in a few special technical problems, and cannot be reduced to a set of rules. The design will take into account visual performance; but an approach based upon visual performance alone is no more likely to result in a significant and pleasing appearance than is a study of acoustics and the performance of the ear likely to result in a Bach fugue. But a design whose aim is aesthetic† is likely to fulfil many of the visual principles instinctively, inasmuch as visual principles are the result of studies of characteristics which are visually inefficient or uncomfortable. Furthermore, the designer must be willing and competent to arrive at an aesthetically pleasing and suitable concept as an exercise in aesthetics with, initially, no principal reference to engineering; though he may modify the design later for engineering reasons, and engineering considerations will be constantly at his elbow, even though aesthetic requirements take the lead. If the aesthetic aim is clearly defined at the outset, the designer can use with confidence sophisticated engineering means, knowing why he is doing so, which would be unjustified in the absence of a clear aim. Thus good aesthetics lead to better engineering. Aims must precede means. A valuable rule is: 'Decide first how you want this interior to appear; and only then engineer it, using the most stable and simplest means available'—simplest, because simple means are likely to remain serviceable. The reverse approach, always a temptation, to think first of what lighting equipment or system to use, perhaps with an eye to a favourite fitting, or one seen elsewhere, is a recipe for disaster.

Basic Questions

The designer in approaching a new project will proceed by certain basic questions, e.g.

1. What is this space? Who uses it? When? What for?
2. What do they need to look at particularly, when using the interior? That is: what are the 'principal objects of regard'? Where from? Are there secondary objects of regard for users, or those important to, say, visitors rather than to users?
3. What is the 'mood' appropriate to the interior? Is there more than one, for different uses? For example a rest room or lounge, occasionally used for concerts or lectures.

† The aesthetic aims must be those which will be deemed pleasing and suitable by the majority of users; not necessarily those which please the architect. 'Avant-garde' ideas are likely to please only their creator and are not likely to fulfil visual principles.

4. What is the present scheme of decoration? Can it be modified?
5. What should be the 'impact view'—the first impression which users receive as they enter? The impact view is almost ineradicable and is very important. Where is the entrance, or entrances? Where will users have come from, and what is its appearance and 'mood'?
6. What should be the changing experiences of users as they traverse successive spaces? What should be the vista if successive spaces can be seen together? Can this vista be reconciled with the requirements for each space considered in isolation?
7. What should be the appearance of people in the interior? People are the most important furniture; their presence significantly alters its appearance and even the impression of its scale. Will there be a 'performer/audience' situation, as in a lecture theatre or a church where the appearance of the speaker is important to the audience—even in hearing him, since we all lip-read—and the appearance of the audience is important to the speaker? Will their appearances be appropriate? It is very important to consider not merely the empty interior, when the architecture and the furniture are all there is to appreciate and may carry undue weight; but to consider it occupied and used, for then the room becomes the setting for the activity of the people in it, and the architecture and furniture recede. Characteristics criticised in the empty interior may become quite unimportant when people are filling and using it.
8. Notwithstanding the last comment, will the appearance collaborate with the architecture and decoration? It is important to be sensitive to them; and it may be possible to assist them, e.g. by reducing the impact of an unavoidable architectural blemish, or distracting attention from it. Are modifications to the decoration permitted? Can the desired appearance be 'painted on'? Can the decoration required be provided, and—most important—maintained? It is usually impracticable to specify individual decoration for particular interiors in a large building.
9. Will the interior be used for more than one activity? Will they require different appearances, and if so, what?
10. Will different lighting conditions have to be considered, e.g. daylight and artificial light? Will the same Designed Appearance be required for each? Will decorations envisaged for one suit the other also? Will the change affect objects of regard and 'mood'? Will the 'focus' of the room be altered? For example in a living room the focus by artificial light is often the fireplace or the TV set; by day it

may be the window and the view. Will the two systems affect modelling? They generally do.

11. How far should appearance in artificial light seek to imitate daylight effects? For example are there features or important details designed to respond to daylight from the side windows, as in a Gothic church? Are the daylight effects worthy of imitation? Often they are not; windows can seldom be sited ideally.

12. How far should designed appearance use stage or TV effects? There is much to be learned from them; but they are designed for the audience or for cameras which are usually outside the space, not for people living in the set: they are frequently acutely glaring to actors, who have learned to accept it, but would never be tolerated in a normal interior. Nevertheless stage or studio equipment may often be effectively used, provided that it lights only features which are visible by users but inaccessible to them, so that glare is not caused, and provided that the equipment can be adequately concealed and that its settings can be locked.

13. Will the appearance of the interior be obviously purposeful and linked to its use, in a coherent way? It is the interior and the activities in it which should catch the attention, not the lighting; and moreover the interior in its proper use. Lighting equipment should be as inconspicuous as possible except when it is deliberately used as an embellishment. It is sometimes justified to provide lighting fittings as embellishments, of low power, and additional concealed equipment to provide the lighting. Striking lighting effects may be used when they are appropriate to the use of the room; but if they are not, e.g. if they emphasise features which though interesting are not objects of regard arising from the use of the room, or if they are mere 'doodles in light', they produce a distracting or theatrical effect rather than one which is naturally dramatic. The appearance to be sought is one in which users find themselves looking quite naturally at the features important to its proper use, and finding them appearing pleasing and appropriate, without their attention being drawn to the lighting equipment or even to the fact that conscious design has been applied to the appearance. The best compliment to a Designed Appearance scheme is that it should not be noticed, in an interior which appears pleasing and appropriate.

As the answers to these questions accumulate, a suitable appearance emerges in the designer's mind.

TECHNICAL MEANS FOR ACHIEVING DESIGNED APPEARANCE

The vague language, characteristic of aesthetics, hitherto used must now be translated into precise technical requirements. The problem occurs in other disciplines; André Previn recounts amusingly how he once asked his principal horn player that his solo passage should be 'more vibrant; more compelling; I want it to come forward ...' and so forth. The player replied: 'Do you want it louder, Mr Previn?' Which, the conductor confessed on reflection, was precisely what he did mean. We have to achieve similar precision.

Technical Means at the Designer's Disposal

The lighting designer has a rather limited set of technical means at his disposal; his aesthetic concepts must now be expressed in terms of them in a strictly practical way.

Pattern of Luminosity

The level of luminosity, and its pattern, are the principal means of achieving the 'mood' of the interior, and of directing attention to appropriate features.[2] Attention is attracted to the brightest features, and distracted from the darker ones. If the range of luminosity is small the scene lacks attraction or focus and appears dull; if in addition the level is low it will appear gloomy. If the level is high but the range small, the effect though bright is still dull and uninteresting, for there is little emphasis or modelling. A high interior lit by pendant equipment at a fairly low height which does not light the upper parts appears low and gloomy and its height is concealed; conversely, indirect light alone directed to the ceiling, and some types of overall translucent ceiling result in the ceiling becoming the brightest feature in the room and claiming the attention; the room looks dull and inverted.

Some dark regions are necessary for an attractive appearance as well as bright ones; but they must be in the right places. The bright features must be those of most significance to users; the darker ones those of little importance or those from which for some reason we wish to distract attention. Such dark features, provided that they are of little intrinsic interest, are welcomed as 'rest centres'. But if a region is left dark which is important to some users, they will resent it and may complain. If the range of luminance becomes large the effect becomes exciting, bizarre and theatrical, with very strong modelling—excellent if that is appropriate. Thus the 'mood' of the interior—restful or invigorating, relaxing or stimulating—is controlled largely by the pattern of luminosity; the pattern will be arrived at intuitively, but it is by the pattern that the aesthetic experiences can be defined and transmitted.

The pattern can affect the apparent form of the interior. If the sides are left dark, as they may be in the aisles of a church, the building appears narrower or even cramped; conversely, spaces at the sides which appear bright make the interior appear wider and may invite users into them, e.g. bookstacks at the sides of a reference library. Control of luminosity can thus be deliberately used to correct visual faults or can make the room appear wider or narrower, higher or lower.

Decoration

The luminosity discussed above is the impression of lightness of a surface or feature. It is subjective, not immediately measurable; it arises from the luminance compared with the adaptation level of the observer. The luminance of a matt feature is the product of the illuminance which it receives, and its reflection factor. The decoration of an interior may be more potent in determining its appearance than light can be; it is sometimes possible to 'paint on' the appearance, so avoiding the need for sophisticated control of light; provided always that the decoration is maintained unchanged. A given luminance may be provided by a light decoration moderately lighted, or by dark decoration strongly lighted; however, the eye is not deceived by the change except in special circumstances. Attention may be caught by a dark feature, say a vase, strongly lit; but the viewer knows perfectly well that it is a dark vase and not a light one. But if a dark picture is lit by a profile spotlight *which does not light its frame*, the layman is not aware of the extra light and the result is most striking.

Decorations may determine the 'mood' of an interior to a degree which cannot be overcome by the lighting. Dark decoration not only makes the interior look gloomy and often glaring; it also affects the conspicuousness of objects in the room, especially people, whose faces will appear light on a dark background; modelling is increased sometimes to an unacceptable degree. Downlights in a low dark interior make inevitable very strong modelling and grotesque facial shadows. Conversely, light coloured decorations and furniture throughout lead to a flat appearance with little relief (except for such features as clothing) in which dark objects, e.g. a dark pair of gloves in a shop so lit, are quite hard to see. One carries them about vainly seeking 'a better light'. They appear in all positions with a brightness near the threshold of the local visual range.

Colour

Colour can be used very subtly, in the hands of an expert, to contribute to the desired appearance of an interior; but it can be disastrous in inexpert hands. It is the colour of decorations which is the most important, revealed by approximately white light. It is sometimes possible, but not very practicable, to produce desired effects by the use of coloured light, in regions of an interior which are to be picked out and emphasised, such as the chancel of a church or a memorial chapel. This may succeed, provided that the lamps are not interchangeable with the rest of the lighting, e.g. a small region lit by fluorescent lamps where the general lighting is by tungsten lamps. It is most unwise to rely upon, say, fluorescent lamps of a particular colour, for they are almost certain to be mixed up during maintenance. Colour filters can be used on projectors; but they may fade if of plastic or crack if of glass, or be left off. In any case coloured light floods the scene and destroys colour contrasts in the decorations.

Colour decorations can be seriously altered by the spectral properties of various 'white' lights; this occurs most markedly with 'pastel' shades of low saturation. Where colour discrimination is essential, as in art galleries, in some problems of inspection, and for clinical purposes, lamps giving particularly good colour rendering are available; but their efficacy is low. Even in these critical situations the effects may be less serious than is often feared, but care should be taken; advice is given in relevant CIBS publications. There are also strong effects of inter-reflection, to be found in small, high rooms decorated with pastel colours up to the ceiling. This may greatly and unacceptably increase the apparent saturation of a delicate pastel colour. The appearance of a given decoration colour may vary markedly in different parts of the room, because the relative contributions of direct and inter-reflected light, of different colours, vary. In daylight the light from the window to various surfaces varies in colour as it comes from the sky, from external buildings, or from grass.

If an interior is lit throughout by lamps of the same colour, the effects of that light on the colour appearance of features in the room—which may be considerable—diminishes, as observers adapt to the colour. Objects known to be white will appear white, even if lit by yellowish light, e.g. by filament lamps; the observer's 'white point' moves into the yellow, and an accidental spot of sunlight appears markedly blue. If a restaurant is decorated in soft blue, the diners' white point moves into the blue. As a result of this, and also due to the effects of simultaneous contrast, reddish meat will appear more succulent and complexions are enhanced. However if the restaurant is decorated in a pink shade, people appear paler and meat less appetising.

Much more important than effects which observers have adapted to are the transient effects to which they have not, especially when they have come from an interior with one scheme of decorations and light into a different one. Colour changes can then appear marked and sometimes so unfavourable as to result in vigorous complaints. In all uses of colour, the important effect is the contrast of colours rather than the appearance of one uniform colour. There are strong effects of simultaneous contrast, whereby the effect of a given colour depends upon the colours with which it is seen. One room decorated throughout, including the ceiling, in a pale pink gave no impression of pink until the picture rail was painted white, when the walls and ceiling looked markedly pink. Pastel colours need a true white feature to give a point of reference, or they may have little effect.

Modelling and Shadow

Modelling is the revealing of three-dimensional form by the use of incident light having a marked directional component.[2] It occurs because surfaces normal to the directional component receive greater illuminance than those turned away from it. If an illuminated object has re-entrants, there will in addition be shadows, formed when a projecting part cuts off light from a strong source from another part of the object; these shadows play an important part in displaying the shape of the object.

The degree of modelling and shadows depends on the luminances in the whole field surrounding the object, including the light sources; this is an important concept developed in the section on specifying Designed Appearance. If a small plane surface sited at some point of interest in the interior is turned about into different orientations, its illuminance will in general vary; if matt, its luminance will vary proportionately. If it is lit by a small source in an otherwise dark room, its illuminance will follow a cosine law. But in habitable spaces, as it turns away from one source it usually turns towards another, direct or indirect, and the range of illuminance is reduced. This is why there is so little modelling in interiors which are well lit and have light decorations throughout.

A shadow is a region within which a light source is obstructed by some intervening object. The density and the colour of the shadow depend upon what sources are left when that one is cut off; the sharpness of the shadow depends on the subtense of the source at the intervening object, and the distance of the latter from the cast shadow. Small sources produce sharp shadows; large sources produce soft-edged shadows, sometimes so soft that they are not discerned as such. If several small bright sources light an object, an obstruction will produce several corresponding shadows; but they will be less dense when only one source is cut off at a time. Such shadows can produce interesting (or maddening) patterns. A line source produces sharp shadows from an opaque edge parallel to its length, but very soft shadows if the lamp and the edge are relatively inclined.

Modelling and shadows can to some extent be controlled independently;[2] a large fairly bright source at one side of a room, such as a strongly lit wall, will produce good modelling but little discernible shadow. Two strong small modelling lights on each side of an object will produce little modelling, for their effects cancel, but two sharp and often confusing shadows. For modelling of architecture to be effective there must be sufficient elements of the architecture or content which have interesting or significant solid shapes, and features must be matt or nearly so (see section on non-matt surfaces). Preferred modelling, in England, arises from light with a horizontal component reaching objects obliquely from the line of sight; if such light is provided throughout an interior it tends to unacceptable glare, for some observers will be looking into the modelling light, though in practice laymen using the interior for its proper purpose may not be troubled by it. However, it is often possible to avoid the difficulty by confining deliberate modelling to important objects of regard inaccessible to users, such as the altar in a church or the platform of a concert hall, or a pipe organ. This produces an impression of modelling which is 'referred' by observers to the whole interior. However, interiors of, for example, Gothic style which respond in all parts to directional light from side windows may demand a good deal of modelling throughout.

Modelling depends not only on the direction of incidence on an object, but also on the direction of view. If these two directions differ by preferably about 45°–60° on plan, or in some circumstances up to 120° a face will appear attractively modelled; if it is seen 'down light' it will appear flat, as do many flash photographs. If seen 'up light', i.e. against the direction of modelling and back-lit, modelling will almost disappear and the object will be seen in silhouette with a bright outline, which detaches it from its background. In a building the modelling direction must be planned for the most important viewing direction, and should be coherent—in the same direction—throughout the space, or it may appear incongruous.

Appearance of People

People form the most significant features in any interior; their appearance is important. Even a stark modern interior, the fabric of which in itself provides little opportunity for modelling, may profit from some modelling light to enhance the appearance of people. Those engaged in significant activities, e.g. speakers or performers, should be carefully modelled, even if they experience some glare—they may well not notice it. An exception is the player in an orchestra; if when he looks at the conductor he is faced by a powerful modelling spot almost in line with the conductor's head, he experiences strong after-images and can no longer read his music.

'Punctuation Marks'
Very large buildings such as office blocks are not usually suitable for sophisticated lighting design for reasons discussed above. They tend to uniformity and to almost frightening impersonality and monotony, and it is easy to lose one's way. This may be alleviated by the provision, here and there, of 'punctuation marks' such as a piece of statuary, a flower display or a mural, carefully lit and, where appropriate, modelled, which will form landmarks and points of interest to users. At Sydney Opera House the staff passages are mostly curved and uniformly decorated; there is little sense of location. To give some break in their uniformity the architect asked for deliberately non-uniform lighting to give them some interest.

Non-matt surfaces
Hitherto surfaces have been assumed to be matt or nearly so, their luminance being independent of viewing direction and equal to the product of their illuminance and their reflection factor. However in practice many important surfaces are not matt. Specular (mirrored) surfaces behave entirely differently from matt surfaces; they form optical images of objects in the interior and of light sources 'visible' to them, images which are grossly distorted unless the surface is plane. The luminance of these images is proportional to the product of the *luminance of the object reflected* and the (specular) reflection factor of the mirror, and independent of the illuminance incident on the mirror.

A mirror surface cannot be made brighter by illuminating it; all that happens is the creation of further images. Some decorative elements may be specular, and demand special care. The organ pipes in the Concert Hall of Sydney Opera House are specular; to reveal them and to give 'life' to the organ—an important object of regard—a long vertical continuous line of fluorescent fittings was provided at one side, screened from all parts except the organ pipes; each pipe formed a bright line image extending for its whole length, indicating its form and finish. The position of the viewer is important; it was impracticable to provide a reflection for the whole length of every pipe from every seat; but sufficient was achieved to emphasise and model the organ adequately. Another example is often found in a church when the altar cross is of plane polished silver or brass. Viewers will see only what the cross reflects, usually something at the back of the church; the cross itself may vanish. In Durham Cathedral the high altar cross stands on a wooden wedge to tilt it back slightly and to give, at least from some positions, a reflection of the west window in daylight. The effects of disappearance can be exploited; at the Sydney Opera House small downlights were required between large concrete beams; the architect wanted them to be concealed. They could not be recessed. One solution (not eventually used) was to make them square on plan with a mirror finish, so that they merely reflected another part of the concrete, and became almost invisible.

Reflections in the glazing of windows can be troublesome, particularly in an art gallery showing glazed pictures. It is very difficult to avoid in daylight reflections either of windows or of spectators, and architectural design may be seriously affected in order to avoid them. Reflections of artificial light sources can be more easily evaded but careful attention must be given. Reflections of sky in window glazing whether looking out, as at Sydney, or in, as in a show window, can be eliminated by so curving or inclining the glass that the viewer sees reflected in it only a black surface provided. At Sydney, viewing panels were inclined outwards to reflect a black louvre set above them, so that viewers were untroubled by reflections of overhead glazing by day or of themselves at night.

Many surfaces are not simply matt or specular: some, e.g. glazed ceramics, exhibit matt and specular reflection at the same time. Many others have properties intermediate between matt and specular. At the position where a specular surface would form an image they form a bright, extended 'hot spot' the luminance of which is proportional to the illuminance at the surface. The luminance of the centre of the 'hot-spot' is less bright than a specular image would have been, but much brighter than a matt surface of the same reflection factor would have been. With all such surfaces the directions of incident light and of view affect the luminance of the object. Reflections are more difficult to avoid than with specular surfaces, because the hot spot is large, and persists at angles at which a specular image would have been avoided. This is important in, e.g. picture galleries especially with dark unglazed but varnished pictures.

Occasionally one encounters anisotropic surfaces, the properties of which alter if a specimen is rotated in its own plane. The most common are woven damask patterns, in which shiny threads run in directions mutually at right angles in the pattern and the background. The threads reflect in directions at right angles to their length. They are sometimes found in churches as altar cloths, and with certain directions of illumination and view the pattern entirely disappears. Suitable directions must be found by trial. A similar effect can occur with a dark oil painting in which the artist has used a stiff brush making strokes in several directions; such a canvas is very difficult to light, for one direction or another will shine and spoil the dark regions of the painting.

SPECIFYING DESIGNED APPEARANCE

The designer when he has formed his concept of Designed Appearance must communicate it to his client, and specify it to the contractor. He must express in conventional technical terms the equipment to be used, its sites, its adjustments, and electrical loads.

His concept may and usually should be expressed by perspective drawings of the interior as seen from one or more critical viewing positions; they should be accurate and drawn for a normal viewing distance, and shaded to indicate patterns of luminosity, and modelling. Such a drawing can be understood by the client and the architect; it is also a valuable discipline to the designer, who must consider *every* feature in view if he is to shade it correctly. This conveys the appearance which he has evolved. He now needs a bridge between his concept of appearance and the conventional engineering terms. This bridge exists in scales of Apparent Brightness, or Luminosity, expressed numerically on an empirical scale and related to luminances in the view and to adaptation level. Such curves have been published by several workers; an early example is shown in Fig. 1.[1,2] Their theoretical basis and their accuracy have been challenged; but in practice these objections have little practical significance to the designer, for whom consistency is more necessary than precision. By estimating luminosity in a number of situations it is possible to prepare a scale ascribing describable effects to different levels. Each system for numerating luminosity will give rise to its own scale; in practice it is sufficient to use one system and keep to it; for the precision required is not usually high. The concept of the design is more important than the numerical precision with which it is achieved, which is fortunate; limitations in admissible sites and available lamps and equipment preclude a precise solution.

A television set provides an analogy. The user can by controls vary over wide limits the brightness, contrast and colour of the received picture, and each user has his preferred settings. So long as a well-composed picture is received, it will remain acceptable over quite a wide range of settings. But a

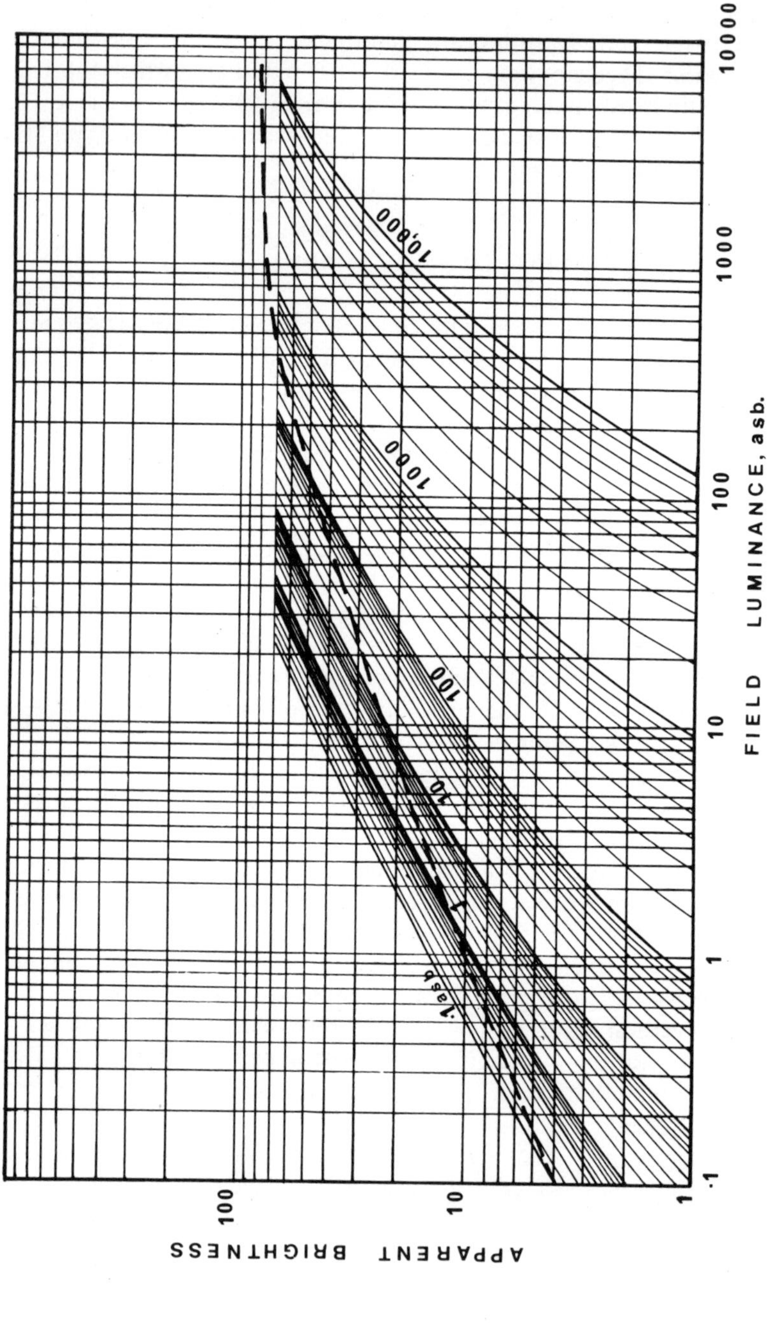

FIG. 1. Apparent brightness curves.

badly composed picture cannot be improved by any amount of adjustments. If in a Designed Appearance scheme it has been decided for good reason that feature A should appear brighter than feature B, that is what matters most; the amount by which A exceeds B is not usually critical. However, if they are reversed, with B brighter than A, the effect may be deleterious.

Luminosity and Adaptation

Using the scale of luminosity (Fig. 2) as a guide, the designer can ascribe to each main element in the perspective a number expressing luminosity: Fig. 3 shows a simple case. He makes these ascriptions intuitively, to numerate the appearance he wishes to achieve. The set of curves, Fig. 1, permits him to transform these numbers into values of luminance, provided that he can estimate the adaptation level. He can usually do this by inspection of the

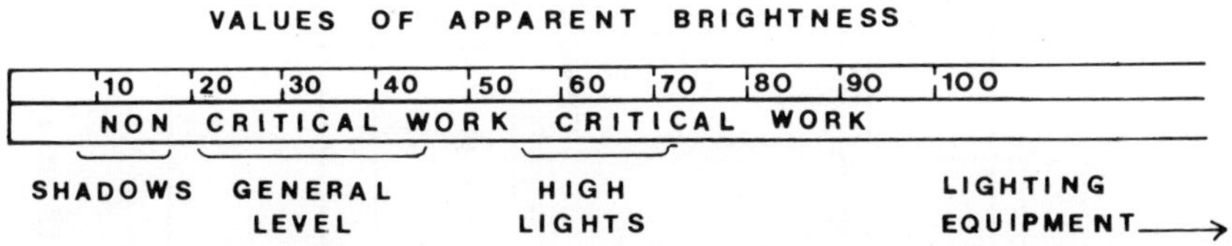

FIG. 2. Scale of apparent brightness and some typical values.

luminosity field in the perspective, and by estimating a mean value. Using this value with the dashed curve in Fig. 1, which shows the luminance of the adaptation level, he finds the adaptation luminance, and also the curve from the set which is to be used to convert field luminosities to luminances. This operation is fairly straightforward with concealed sources, but estimates of adaptation level become progressively more difficult with exposed luminaires, with windows, and with large areas of sky. Here more work is needed; the concepts of adaptation and of glare merge.†

† The presence of bright light sources in a scene not only induces a haze veil (leading to disability glare) and increases the luminance range (leading to discomfort glare) but also, by raising the adaptation level, depresses the apparent brightness of objects of regard. (Shielding a bright source with the hand, usually employed as a demonstration of glare effects, improves 'visibility' because the apparent brightness of the objects of regard increases.) This decrease underlies and probably forms part of effects attributed to both discomfort and disability glare.

At present we do not know how to determine the effective adaptation level in a scene which includes bright artificial light sources, or a bright sky. In some problems, e.g. for an observer about to enter a vehicular tunnel in bright daylight, the visual problems cannot be numerated for lack of further data on adaptation.

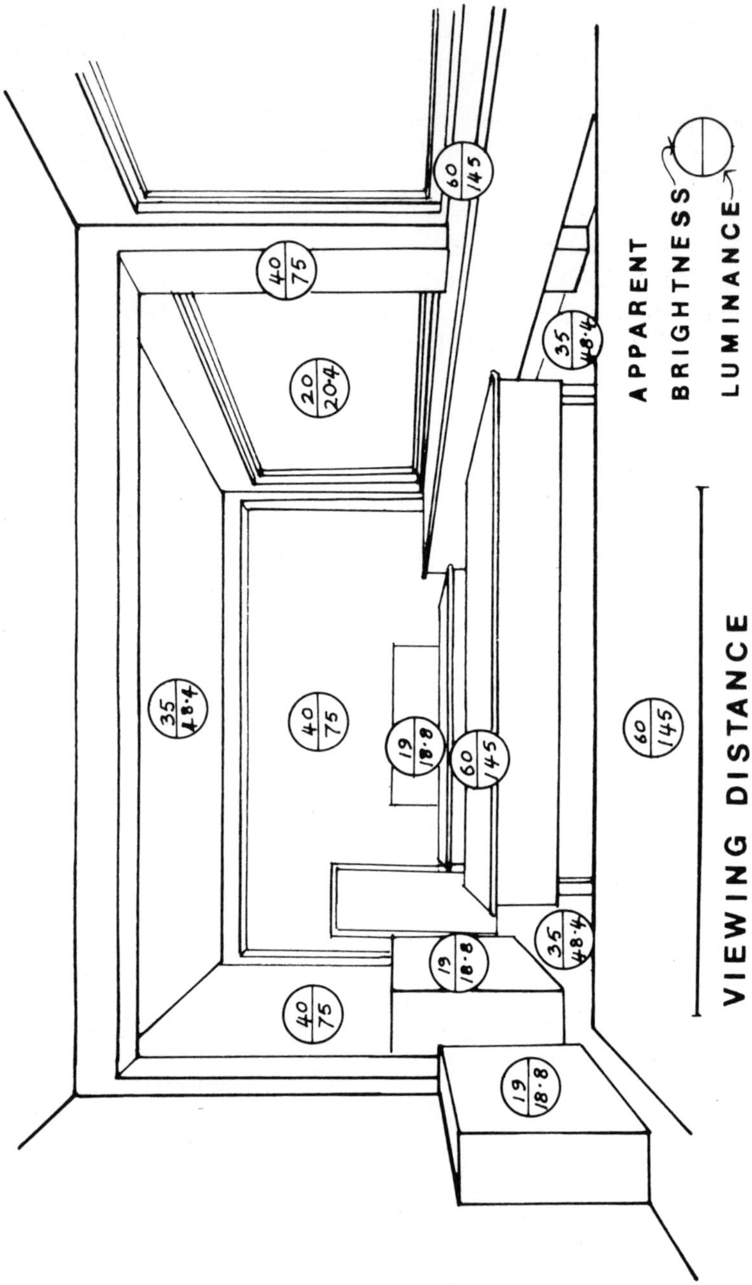

FIG. 3. The design of the visual field in a small light assembly shop.

The perspective with luminances also marked on it (see Fig. 3) is the main base for visual and engineering design. It is a statement of the luminance field in the eventual installation, and it is the basic concept of the lighting of interiors, from which flow all the other concepts. The interior behaves as a 'sky' surrounding any point within it, and this sky, comprising the principal surfaces,—walls, floor and ceiling, furniture, and visible luminaires—brings about the luminance of any object, its illuminance, modelling and reflections and shadows, and visual effects such as adaptation, visual performance, discomfort or glare, and aesthetic concepts. The principal items on this list we can now calculate.

Illuminances

From the prescribed luminance of each main surface, and its reflection factor, comes the total illuminance which that surface should receive. That illuminance has two components: the illuminance received directly from the luminaires, and illuminance received from the rest of the 'sky'—the other main surfaces. If we can estimate the latter, and subtract it from the totals required, the remainder is the direct illuminance required to each main surface from the luminaires, which is the basis for the installation design.

Inter-reflection Illuminance

In a rectangular enclosure each main surface faces five others, each of which acts like a window or skylight illuminating the sixth. Their luminances are known; the flux which reaches each main surface from one of the others can be found using the transfer factors given in the IES Code;[6] this gives the average illuminance of the sixth surface from each of the other five. A simpler method[3] gives the illuminance at the centre point of each recipient surface, which is if anything more representative. If the building is not a rectangular box, e.g. some churches and concert halls, the whole interior is considered as a 'sky' of variable but known luminance. For a uniform sky the illumination on the diametral plane is given by $E = \pi \,.\, B$ where B is in cd/m^2, or more simply $E = B'$ where B' is in apostilb, which is more convenient. If the 'sky' is represented by a hemisphere, it can be shown that (Fig. 4) one half of the illuminance on the diametral plane is provided by the belt from the horizon to 45° and the other half by the cap from 45° to the zenith. If in either half the luminance varies in azimuth, the mean can be estimated. Using this concept the inter-reflection illuminance can be estimated sufficiently accurately without formal calculation. For surfaces other than the floor, the system is considered rotated through 90° for the

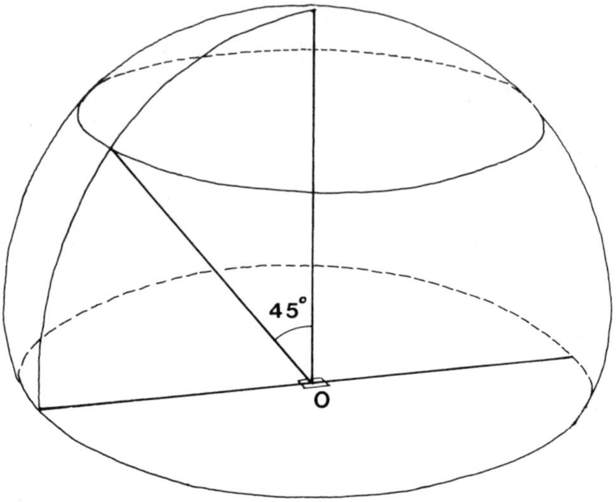

FIG. 4. Horizontal illuminance at 0, from complete hemisphere of luminance $B\,\mathrm{cd/m^2}$, is πB lux. $0{\cdot}5\,\pi B$ lux comes from 0°–45° zone. $0{\cdot}5\,\pi B$ lux comes from 45°–90° zone.

walls and 180° for the ceiling or vault. A more detailed discussion for rectangular rooms is given in reference 3.

The inter-reflection illuminance is usually small compared with the total required, so that its estimate need not be precise. The direct illuminance itself cannot be accurately realised because of practical limitations of available lamps and equipment, and of siting. It sometimes happens that in the upper parts of the room inter-reflection provides all that is required; sometimes more. (See Table 1 and Fig. 3, ceiling.) This can be adjusted either by changing the specified luminosity or by changing the reflection factor.

Calculations

It is now possible to construct a table showing, for the simple case of Fig. 3, the steps of the calculation (Table 1). This table needs scrutiny to make sure that it leads to sensible values and light distributions; some reiteration may be necessary if there are exposed luminaires, which may affect adaptation level. However the calculated values of direct illumination should be taken as guides rather than as exact specifications; a necessary step between the formulation of the aesthetic design and the engineering design which will realise it. Other calculations are of course possible, such as glare using the

TABLE 1

Location	*Apparent brightness (luminosity)*	*Luminance cd/m²*	*Luminance apo-stilb*	*Refl. factor*	*Total illum. lux*	*Interr. illum. lux*	*Dir. illum. lux*
Benches	60	46	145	0·30	485	54	431
Floor	35	15·4	48·4	0·15	323	32	291
Walls	40	24	75	0·40	188	62·6	126
Ceiling	35	15·4	48·4	0·65	73·3	71·2	2·2
Adaptation level	40						

method of the IES Code, and modelling using either the vector/scalar concept discussed in Chapter 1 or the cylindrical concept.

The whole process becomes much simplified with experience and from practical considerations. The most important problem is found to be not in getting light where it is wanted, but in preventing it from going where it is not wanted, This approach will help, especially if the design of the direct lighting includes the useful device of plotting on an isocandle web or on a perspective the view of the interior from the lighting equipment in one bay. The light distribution of the equipment can be superimposed as isocandle lines, so that the contribution to each point of walls or floor can be read off, enabling illuminances to be found immediately with a Holmes slide rule. Spill light in unwanted places can be spotted immediately. These devices need some simple geometrical aids which there is not space to describe here.†

ENGINEERING CONSIDERATIONS

The engineering of the scheme proceeds in the designer's mind simultaneously with the concepts of appearance. As he thinks about modelling and glare, he is looking for sites for equipment which might achieve the one and avoid the other, and also the means of maintaining equipment in these positions. These questions, together with stability of performance, are vital. It is pointless to work out a sophisticated design unless the system will be working as it was intended after ten years.

Design is easier with concealed equipment than with exposed, and also the installation is much cheaper. At an early stage in the design the sites for

† The stereographic projection, applied by J. A. Lynes[7] to floodlighting problems, if extended to a complete hemisphere, includes these aids and greatly simplifies the procedure.

equipment will have been found and means of reaching them thought out; the total electric load at each site and some idea of switching and control will have emerged, so that provisional estimates and feasibility studies can proceed. If the scheme is accepted at this stage, more detailed design can proceed. It will be useful in a large building to arrange a trial of say two bays, mostly for checking the design. If possible the client should be discouraged from being present, for he is liable to be misled; it is difficult even for the expert to estimate the final result from a small trial. Experiments with the object of *establishing* the designed appearance are not however useful. Such experiments are almost impossible to arrange in a building under construction; they cannot be done until the interior is virtually finished, decorated and furnished, and then it is too late. It is also very expensive.

Stability and Maintenance

The great importance of these has been stressed. The installation must be capable of maintenance by staff not there at its inception, and they must be able to maintain it simply and safely. The lighting equipment must be very well designed, simple in concept (complex equipment is seldom stable), capable of being securely locked in position but simple to relamp and clean without disturbing its settings, and not liable to soil easily or to corrode. Correct lamps must be used for replacement; ideally, for concealed equipment, stored near them. If staff have climbed up to the roof with a wrong lamp, they are unlikely to go back for the right one. Maintenance must be safe; proper walkways, railings and ladder hooks must be provided. Large lamps need both hands to insert them. It must be impossible for a part being handled, a tool or a lamp, or cracked glassware, to fall to the floor below. Each luminaire must be unplugged at the site, both for safety and to let it get cool enough to handle. Lighting ports must not be glazed, for the glass quickly soils and may crack. Open ports, however, may admit draughts and spoil acoustics, so that fittings may have to be boxed in, with only enough ventilation to prevent overheating. If projectors have to be aimed, the man at the fitting should be able to see what he is doing; two men may be needed, with radio communication. The aiming adjustments must be smooth and simple to operate without sticking or jamming, or the need to handle a hot part.

Choice of Lamps

Lamps specified for Designed Appearance schemes should always be of established types likely to remain on the market for a long period, and a good stock held. The choice of lamp turns on the lamp price, life, efficacy, source size, available outputs, and on the hours of use per year in the installation. If illuminance levels are high and burning hours long, and the light control not critical, fluorescent lamps of low working cost and long life are an obvious choice. For an interior used only a few hours per week, but where light control is critical, tungsten lamps are justifiable, especially PAR reflector lamps which are small and very stable, and tungsten halogen lamps. Filament lamps have a short life and low efficacy; but if for other reasons dimmers are desirable, under-running of the lamps will greatly increase their life.

Electrical Installation and Switching

The circuits and switches should permit maximum advantage to be taken of the installation, especially for buildings having more than one use. A cathedral provides many possibilities, with many different services in various places, and other activities, such as concerts, for all of which there should be appropriate and easily switched lighting, providing not only for the occasion itself but also for access and safety. Switching should permit several different levels of lighting. The lighting scheme should provide more than one unit to light each feature, so that a single lamp failure does not leave a dark patch; important objects of regard should for the same reason be lit by at least three units. These can be switched on two circuits to give three levels. Schedules of switching for each type of occasion should be prepared, one copy being by the control point. It is a great advantage if the operator can see the effects of his switching. It is also most important that switching should be in the hands of a competent person who will not simply turn everything on on every occasion.

EXAMPLE: ELY CATHEDRAL

Ely Cathedral was lit throughout by the method of Designed Appearance.† The building is very long, narrow and high, consisting of a Norman nave and transepts, all with aisles, a wide triforium and clerestory with access. There is a mansard roof to the nave, of painted wood, exposed trusses in the transepts, and a groined stone vault in two styles in the chancel, which is a little lower than the nave. The crossing is the unique feature, formed as an octagon and surmounted by a beautiful timber vault supporting a large

† Further details and photographs are given in references 4 and 5.

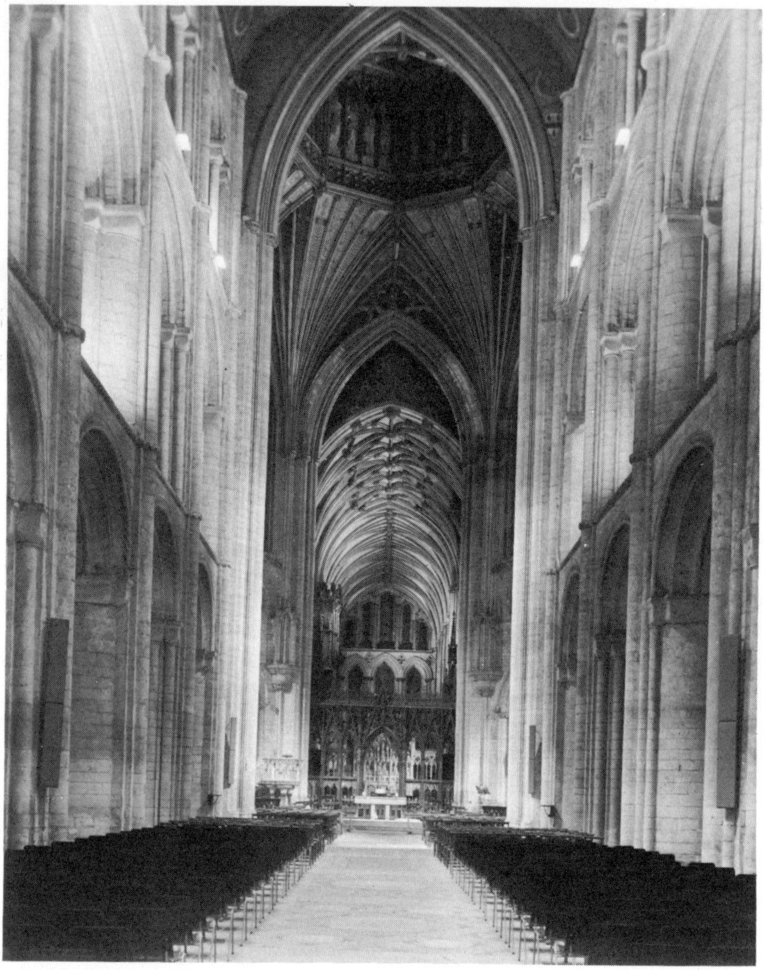

FIG. 5. Ely Cathedral by night. (Courtesy Osram (GEC) Ltd.)

lantern admitting daylight. The nave altar is set below the lantern, and many services and musical occasions take place there. The photograph in Fig. 5 shows part of the nave, the crossing, and the screen and chancel beyond, by artificial light.

Designed Appearance

As in any church, the appearance of the building contributes to worship within it, and is specially important. Lighting levels need not be high but not only must there be adequate illuminance for access and for reading, but the objects of regard important to worship, including the officiating clergy, must be appropriately emphasised and modelled. More light should be provided in the Octagon, where many services take place, and in the chancel. The equipment should obviously be substantially concealed, for visible fittings would form distracting lines of lights. Light must be provided downwards to the floor and seats over the whole of the main spaces; further light should reveal and if possible model the arcades and the vaults so that their full height can be seen, without making them so bright as to claim attention. The main objects of regard should be emphasised and modelled—the high altar in the chancel, nave altar and pulpit in the crossing; the octagon vaulting and if possible the lantern should be appropriately modelled. Calculation of apparent brightness, luminances, glare and modelling were made on the lines indicated above.

Methods of Lighting

Main Spaces

The methods of lighting are shown in Fig. 6, a perspective view showing the octagon, nave altar and pulpit, north transept and part of the nave. The main spaces were all lit in the same way.

The figure is drawn in perspective, to indicate the complex three-dimensional space in a single diagram; it has necessarily been drawn for a very short viewing distance and an unusual viewpoint, and consequently appears to be distorted when viewed normally. If, however, it is viewed from a point opposite to the cross marked in the nave clerestory and about 5 cm from the plane of the paper, and focused by a lens of about 7·5 cm focal length placed at the eye, it will be found to give a remarkable impression of reality, scale and height. The eye and the lens must be turned together to see the margins, which is not easy; consequently such a perspective is useless for conveying aesthetic proposals to a client, for which purpose a viewing distance of about 30 cm and a correspondingly reduced field of view are necessary.

Downward light was found impracticable with concealed fittings. It was provided from the clerestory by specially designed fittings, not completely concealed but very inconspicuous. One is shown in the south transept in Fig. 6 (marked A). The fitting rests on the parapet, with a 'nose' projecting about 18 cm. Three 150W PAR 38 lamps are mounted horizontally in the body, the light from which is projected downwards by a plane metal mirror in the 'nose' to light the floor, the lower parts of the opposite columns, and between them the aisles, for the aisles provided no practicable sites. A

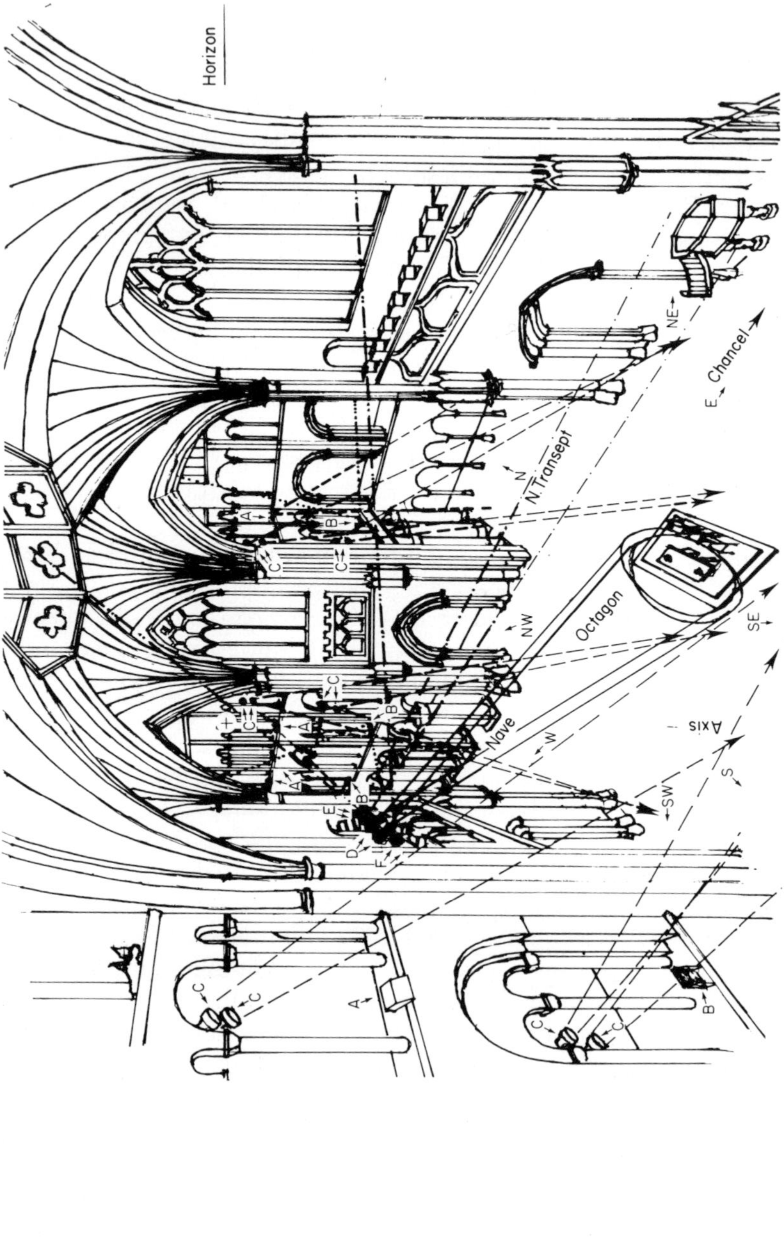

FIG. 6. Ely Cathedral: octagon, nave and transepts, showing positions and aiming of lighting equipment.

louvre cuts out glare; there is no bright splash of light on the face of the parapet below the fitting. These units were costly, being specials, but there was nothing suitable on the market. They were used one per bay on each side except in the chancel where there were two per bay.

Upward light to the upper parts of the columns and to the vault was provided by small standard fan beam projectors with 500W tungsten halogen lamps set on the triforium floor, not visible from the nave floor, in alternate bays on each side, (marked B). An expanded metal screen cut down stray light to the adjacent arch; this was insufficient and an opaque screen would have been better.

The 'nose' fittings were wired on two circuits to give three levels. Modelling of the columns was considered but not provided, owing to difficulties; the main columns were so large that they hardly needed it.

Octagon

There were no readily available sites in the octagon for equipment to light the floor; downward light was provided by four groups of four 300W Par 56 reflector lamps (C on drawing) set as shown in the eastern bay of the nave on each side (only one is visible in Fig. 6) and in the adjacent bays of the north and south transepts on the west side, lighting the Octagon floor through the main arches. These units were not visible from the nave. They cause some distraction to persons sitting in the Octagon but that was probably inevitable.

Objects of Regard

The nave altar, the pulpit and the Octagon vault were all lit by projectors in the southwest Octagon gallery, which provides a suitable site with good access; this gave an appropriate and coherent direction of modelling and grouped the equipment conveniently together. The nave altar is lit by two 500W profile stage spots marked D in Fig. 6, set to cut off light from the three stalls behind the altar in order to avoid glare to their occupants. The pulpit and the preacher are lit by two 500W fresnel stage spots, marked E. Two similar projectors (F), set below the gallery parapet, light the vault directionally and reveal its shapes. It was intended to light the lantern by local units strongly from the southwest, but fire precautions forbade the use of electrical equipment in the lantern space.

The high altar and the space about it were lit by three projectors in the south choir triforium, permitting three lighting levels. The south end of the screen not visible from the triforium was lit by 150W PAR 38 lamps on an adjacent column. This gives strong emphasis and modelling to the carved reredos, and also to clergy. The three east lancet windows were also lit and modelled from the south choir triforium.

The lighting of the cathedral is not without faults, but they are not noticed by the clergy or congregation, and the installation has given satisfaction.

REFERENCES

1. HOPKINSON, R. G., STEVENS, W. R. and WALDRAM, J. M., *Trans. Illum. Engng Soc.* (*London*), 1941, **6** 37.
2. WALDRAM, J. M., *Trans. Illum. Engng Soc.* (*London*), 1954, **19** 95.
3. WALDRAM, J. M., *Trans. Illum. Engng Soc.* (*London*), 1958, **23** 113.
4. WALDRAM, J. M., *Int. Ltg Rev.*, 1975, **26** 87.
5. WALDRAM, J. M., *Light and Ltg*, 1975, **68** 6.
6. 'IES Code for Interior Lighting', 1977.
7. LYNES, J. A., *Ltg Res. Technol.*, 1971, **3** 58.

Chapter 6

EXTERIOR LIGHTING AND TOWNSCAPE

D. McH. Clark
M.A.P.L.E., F.C.I.B.S.
Chief Electrical Engineer, City of Norwich, UK

SUMMARY

This chapter has been written as an attempt to stimulate engineers, planners and architects to strive for the preservation of towns and cities and also to give more depth of thought to design concepts. It discusses the necessity to use our eyes at all times and to appraise what is good and poor design. Installation techniques which require thought and patience to achieve the maximum in concealment are examined and the contributions of signs and shop lighting to the visual scene are also considered. Finally, the value of floodlighting in the creation of a new dimension during the hours of darkness, and in enhancing the appearance of buildings, parks and gardens for the benefit of the general public and tourists alike is briefly considered.

OUR LUMINOUS ENVIRONMENT

The good Lord gave us light. On a dull day one can feel depressed, but as many people experience a bright and sunny day they can be heard to say 'It's good to be alive'. Thus light is important to our well being and a stimulus to our way of life.

Your eyes are the most used organs of your body. The final judge of any lighting installation is the eye. Use your eyes. Everywhere you go you are surrounded by examples of illuminating engineering—good, bad and indifferent. If your eye tells you that there is something wrong with an installation, try to diagnose the trouble and see whether you can suggest a remedy. On the other hand, if an installation pleases you try to analyse its good points. At all times use your eyes.

That there is a fundamental difference between the science of lighting and other applied sciences cannot be denied, for lighting is experienced subjectively, i.e. it can be assessed only by its effect on human beings through their own eyes. Light and the eye form a partnership in which any change in the former produces a change of visual sensation, so that in order to be able to discuss lighting quantitatively something must be known of the mechanism of seeing.

The impression held by many people that artificial lighting is an inferior substitute for natural lighting can arise only from a failure to grasp the tremendous opportunities which light sources present, a failure which results in unscientific and inartistic lighting installations, and which can be rectified only when it is generally realised that the true function of electric lighting is to supplement daylight as well as to substitute for it, and to create seeing conditions and a visual environment which far surpass the capabilities of natural lighting.

Though an adequate quantity of light is essential, it is certainly not the only criterion in good and comfortable vision. That rather indefinable term 'quality' on which lighting comfort depends is just as important, and is determined largely by the appearance, the brightness or colour of everything in view. Whether an Illuminating Engineer should be primarily an artist or a scientist may be open to question. Certainly he should not be interested only in art or only in engineering—there are far too many examples of installations designed by engineers with no artistic ability, and others by artists guided only by inspiration; in both cases ordinary people may well think that the results might have been better. The ideal, of course, is for science to be applied with art, both aiming to satisfy the eye, for this is the only purpose for which lighting is required. We shall never reach perfection in lighting, but there are signs that we are moving in the right direction.

THE ENGINEER'S CONTRIBUTION

There have been many new town and city developments, shopping precincts and pedestrian areas in Great Britain, and there are many more in course of construction or in the final stages of design. At what point in the design stages is the Electrical and Lighting Engineer invited to make his contribution towards the environment during the many hours of darkness, not forgetting a well-balanced aesthetic approach for the visual scene during the daylight hours? A survey of many town and city developments suggests that the early design teamwork requires improvement in order to achieve a better end-product environmentally. It might be that the architect did not make full use of the specialist consultants and made all the decisions himself, or alternatively the specialist did not possess the design flair necessary to stimulate the architect and convince him that jointly much more can be achieved. If the architect completes his design before he consults the Electrical and Lighting Engineer, many difficulties will be experienced and the project will often be an expensive compromise rather than a contribution towards the original design concept. The architect may decide that he will go it alone and design the lighting etc. Seldom will he achieve the exciting lighting environment which could be complementary to his project and give greater enjoyment to the public. It is therefore essential to concentrate on a team effort to produce a well-designed electrical installation, and integrated lighting can benefit all concerned.

How, then, should the Lighting Engineer design for future developments? My suggestions are as follows:

1. Early consultation with the planner and architect. He should approach the designers even if they do not consult him in the first instance. If he has a lively interest in, and understanding of, lighting design and technology, he will soon excite the designer. The Lighting Engineer can point out lighting techniques which will be complementary to the environment. Teamwork will soon develop with understanding on both sides.
2. He can readily explain the colour rendition of the various light sources, how the correct selection of a fluorescent lamp can enhance the product on display, the illuminated signs, the car park lighting, facades of buildings, etc.
3. He can advise on shop interior lighting, particularly window display areas which will play an important part in the night environment.
4. He can envisage ideas for under-water lighting, display case lighting, and means of avoiding unnecessary lamp columns by use of more subtle lighting from walls and structures.
5. Conventional street lighting lanterns should seldom be necessary. Thought should be given, for example, to the fitting of an optical mirror system into a spherical fitting which could illuminate areas efficiently yet give a pleasant daytime appearance (Fig. 1).

FIG. 1. Spherical lantern for 250/400 W SON lamp. Polished aluminium reflector system gives non-axial distribution.

6. If the Lighting Engineer has experience only of basic public lighting techniques, this will limit his acceptability to the architect who may wish to isolate his development from specific traffic routes. Building elevations should provide large surfaces of low luminance in interesting patterns, and direct lighting must have a decorative function and be gay during daylight hours.
7. Multi-storey car parks have very limited headroom, yet many electrical installations have surface conduits and boxes. This means that the mounting height of the lighting fitting is lower than necessary. The conduits will also corrode and deteriorate on the surface (Fig. 2). All circuit cables and conduits *should be contained within the concrete structure*. The maintenance will be reduced and the visual appearance improved.

FIG. 2. Multi-storey car park. Corrosion of conduit and surface trunking.

8. The Lighting Engineer must be experienced in the principles of air conditioning and air-handling lighting fittings.
9. The early agreement of electrical services and distribution will be a worthwhile exercise and avoid many of the unsightly service runs which are so often a compromise.
10. Emergency lighting services are important, and the untidy duplication of lighting fittings and circuit cables in car parks can be avoided by the use of static invertor units providing 240 V a.c. from d.c. supply.
11. Fire alarms, detectors and sprinklers are necessary requirements, but care must be taken not to cause confusion with a common fire alarm system covering a complete pedestrian precinct.
12. Ensure that a maintenance programme is put into operation and bulk lamp changes organised. The use of 'hours run' meters, coupled with periodic tests on the levels of illuminance, will assist in determining the acceptable lamp change period.
13. It is imperative to check the progress of 'as fitted' drawings to ensure they provide the necessary information so essential for the avoidance of many hours of wasteful labour on tracing cable routes and circuit data after the electrical contractors have vacated the site.

There are numerous overseas projects which would require a lengthy description; a few are worth noting for appraisal by both architects and engineers and for comparison with schemes in Great Britain.

La Defense Project—Paris
On a site about 7 km^2, west of the River Seine, where an attempt was made to solve the problem of more space, fresh air and light for living quarters—an important French concept.

The Woluwe Shopping Centre—Brussels (7 km from city centre)
The full length of the building has a central roofed parade 285 m long, 12 m wide with an 18 m wide patio halfway along. Access is available by moving staircases at either side of the parade via the parking areas. Air conditioning is provided. Also fountains, an aviary with exotic birds, shrubs and plants etc.

de Bijenkorf—Eindhoven (Architect, Gio Ponti, Milan)
A shopping and pedestrian piazza with high quality materials such as stainless steel, marble and mosaics.

FIG. 3. Place Carnot, Lyons. Opal acrylic signs, internally illuminated. Ugly brackets are eliminated. The column passes straight through the signs.

FIG. 4. Traffic signs in Place Carnot. Simple clean design. Tapered non-ferrous columns.

Place Carnot—Lyons
Pedestrian shopping area where signs, traffic signals at road junctions, area lighting, seating, paved areas, amenity area adjacent to fountains, water feature amenity and floodlighting are typically *French* (Figs. 3 to 6).

STREET FURNITURE

In general it can be said that in Great Britain we do not put the lasting quality into many of our civic and shopping centres or pedestrian areas. I mention this because the Planning Advice Sub-Committee of the Multiple

FIG. 5. Place Carnot pedestrian precinct. High quality amenity lighting (MBTL lamps).

FIG. 6. Place Carnot gardens and fountain area. Amenity lighting (MBTL lamps).

Shops Federation of Great Britain inspected shopping centres in Sweden, Denmark, West Germany and Holland about 1970. They did not appear to get the message that good design and quality materials can produce good sales and low depreciation.

The electric lighting in town and city centres should create a night scene which serves the social needs of the community, The basic road lighting alone cannot provide the stimulus for social requirements. It should be a combined effort by commerce and the Local Authority.

Shop window lighting is essential to add colour to the visual scene and simple displays can be effective and economical. Well-designed illuminated signs carefully controlled to avoid discomfort glare are important. The floodlighting of buildings and features will then integrate into the central areas and create a feeling of well-being for all to enjoy. The modest petrol filling station could contribute much more to the environment rather than discomfort glare to the passing motorist. One of the important factors in central areas is the ever increasing number of columns for lighting and traffic signs. Great care must be taken to avoid the common practice of erecting a column or post for every conceivable sign or lighting point. This only provides an untidy clutter which confuses rather than informs. It will take more effort and civic pride to improve the design layout. It is important that a full appraisal of the site conditions is given due consideration.

The Road Traffic Signs and General Directions (Statutory Regulations) 1964, No. 1857 and Revised 1975, No. 1536 specifies signs which are unsuitable for many city and town centres. In narrow roads suitable for single line traffic two large 750 mm 'No Entry' signs are unsightly and frequently damaged due to the restricted space (Fig. 7). So why have the same large sign for both wide traffic routes and very narrow roads? Statutory regulations can be expensive and should be re-appraised for specific areas rather than accepted without question. There are few suitable brackets for wall-mounted signs. A cast alloy bracket would enhance the final layout. The practice in Great Britain is to use a clutter of pole clamps for securing signs etc. to posts. A better design concept seen in Paris and the Netherlands is to fit adjustable sleeves on posts for directional signs—the signs having the central post passing through them with a more subtle arrangement. It must be appreciated that Codes of Practice are guide lines only and should not stifle the design abilities of the Lighting Engineer. If we cannot see beyond the Codes of Practice we will never achieve progress and will become static in thought and mind. Unfortunately many engineers consider exterior lighting as columns and lamps. This is a mundane concept and can destroy the visual environment. A further destruction of the visual

FIG. 7. Unnecessary signs. Clutter in a narrow pedestrian area.

scene is a street lighting system on a traffic route which has many lamp columns, few in line and with long and short bracket arms in a confusing array. A sight only to be seen in Great Britain and not in western Europe.

The following extract from the Civic Trust News Sheet (1) was produced to endeavour to promote a greater consideration for the lighting in conservation areas:

'STREET LIGHTING IN CONSERVATION AREAS. In an important new Circular (Roads 33/72) the Department of the Environment has impressed upon highway authorities the need to ensure that in future new road lighting schemes in conservation areas and in historic towns and villages are designed to harmonise with—and if possible enhance—their surroundings.

'The problem. The Circular follows growing concern about the visual disruption often caused in sensitive environments by the erection of tall columns and the use of low-pressure sodium lamps. However intrinsically well-designed they may be, columns of 25 ft or more installed symmetrically along a village street can ruin its scale and informality and produce an effect of instant urbanisation. Though low-pressure sodium is the most economic light source, its rancid yellow tinge and monochromatic rendering tend to spoil the night-time charm of areas of architectural character.

'There is clear evidence that good street lighting reduces the frequency of injury accidents in darkness by about 30 %, and the Circular stresses that authorities must give priority to safety considerations in road lighting schemes. Normally, therefore, schemes for trunk and principal roads must conform to the appropriate British Standard. However, where these roads pass through historic villages a lower standard in keeping with the character of the village may be accepted. In such cases it will remain a requirement that the new lighting should improve and promote road safety at night.

'Conservation Area Lighting Plans. Lighting in conservation areas deserves special attention, the Circular suggests. Highway and planning authorities are urged to prepare plans for these areas, indicating which roads would benefit from the cost of a white light source, which could be lit by low-pressure sodium without loss of amenity, and which have no

FIG. 8. Lighting in a conservation area: Elm Hill, Norwich. High-pressure sodium lamps in 'Windsor' lanterns. Downward lighting by PAR56 lamps with thyristor dimming.

foreseeable need for road lighting. (Roads in the last category would not necessarily be unlit: parish councils and, in some cases, rural district councils have power to provide 'footway lighting'. Intended primarily for the benefit of pedestrians, this modest scale of lighting may sometimes be more appropriate than full-scale road lighting. However, it is not eligible for an Exchequer grant.)

'Once a conservation area lighting plan has been drawn up it should be agreed with the Regional Controller (Roads and Transportation) of the Department of the Environment. Where agreed, the Department will then accept the white light source on its trunk roads and will meet its share of any extra cost involved in its use on principal roads. Preference will be given to schemes using high-pressure sodium lamps rather than colour corrected mercury. (High-pressure sodium lamps are a comparatively recent innovation and yield a warm white tone.) Similar arrangements will apply in any historic towns and villages not yet designated conservation areas.

'The Circular points out that in both types of area the *daylight* appearance of lighting schemes can be improved by the use of special columns, wall brackets and lanterns (the best combinations available are illustrated in the Design Council's Street Furniture Catalogue) and by changes to the normal mounting, lights and spacing. Here again, subject to agreement with the Department and approval by the Royal Fine Art Commission, the Department is willing to meet either the whole of the extra cost involved (on trunk roads) or its share (on principal roads).

'In some cases where there is a night accident problem, road lighting may not be the appropriate solution. Accidents are often found to cluster around one or two road junctions and it may be sufficient to improve the junction layout and/or install illuminated signs and bollards. Where road lighting is considered essential the scheme should be designed in keeping with the character and scale of the roads and buildings. This will often entail lower mounting heights and greater (and perhaps more irregular) spacing.

'Consultation. As in the past, the Department will be responsible for informing the Royal Fine Art Commission of lighting proposals for trunk and principal roads. In addition, the Circular suggests, highways authorities amd those responsible for road lighting on trunk roads under agency agreements with the Department may find it useful to take account of the views of responsible local interests both before and during the design of schemes.'

Unfortunately this circular has now been withdrawn, notwithstanding the recommendations for street lighting in conservation areas should be given due consideration. The Civic Trust will, no doubt, press for a new circular to ensure due consultation for conservation areas.

Statutory regulations for road markings can be a further problem in conservation areas having very narrow road widths. These roads will be retained for conservation reasons. On a traffic route so narrow that the road width will only permit the passage of single line motor traffic one finds double yellow lines on both sides of the road. The motor vehicle can span both sets of yellow lines. There are many such roads and quite a few are one way only or pedestrian ways. Reappraisal of the road marking and sign requirements for such roads could produce a more sympathetic and economical layout.

BRIGHTNESS

Local planning authorities are given powers to control advertisements by Statutory Instruments 1969, No. 1532, the Town Planning and Country Planning (Control of Advertisements) Regulations 1969, as amended by Statutory Instruments 1972, No. 489, and the Town Planning and Country Planning (Control of Advertisements) (Amendment) Regulations 1972 (obtainable from HMSO). The regulations came into force on 1st January 1970 and were introduced by Circular 96/99 issued by the Department of the Environment. Notes on the considerations to be taken into account, where applying the control of advertisement regulations, were issued by the Ministry of Housing and Local Government in Circular 11/62. In granting consent for the display of illuminated advertisements it would seem that the planning authority has the power to impose additional conditions that could include the maximum luminance of illuminated advertisements to a specified value.

There are two separate concepts of brightness, namely photometric or objective brightness and subjective brightness. The photometric or objective brightness, termed luminance, is a characteristic of the sign, independent of the particular observer, and the surrounding conditions. It is capable of being measured and defined as the amount of light emitted in a given direction per unit area of the sign surface. Generally a sign surface is non-uniform and the luminance will vary across it. The luminance will also vary with the direction of viewing, being a maximum for direct frontal observation. With the high diffusing plastic materials normally used in signs (usually Opal 040 or 050) the luminance remains near the maximum over a fairly wide range of viewing angles, falling off steeply for very oblique viewing. Subjective brightness is the effect of brightness experienced by an observer. It depends on four main factors:

1. Luminance
2. Size
3. Surroundings
4. The observer

The luminance of a sign must be related to its particular location. If a sign is placed in a well-lighted central shopping area it will appear to be less bright than a similar sign installed in a dark country area. The one may be attractive and the other quite offensive, although the signs themselves are identical. The luminance is the same for both signs but the subjective brightness is quite different. While it is subjective brightness with which Local Authorities are mainly concerned, only the luminance is within the control of the sign manufacturer. Most signs remain in constant position for many years and are subject, especially during their early life, to a substantial decline in luminance.

It is general for the planning authority to introduce a system of control that imposes a limit on the maximum luminance of an advertisement, as a condition at the time of granting consent to display such advertisement. The luminance data can be checked by the Local Authority Lighting/Electrical Engineer. The local area should be divided into zones related to the existing lighting, thus providing reasonable control of the sign luminance in comparison to its immediate surroundings. The Association of Public Lighting Engineers recommendations are for four zones. These are as follows:

Zone 1 Recognised display centres, such as Piccadilly Circus—recognised as centres for the display of large scale illuminated advertisements, some of which may not relate to the business of the premises on which they are erected. See Table 1.

Zone 2 Major shopping areas—these are the main shopping streets of towns and cities which have well-lighted window displays and road lighting.

Zone 3 Other shopping areas—these are generally the shopping areas and precincts serving residential areas.

Zone 4 Poorly lighted or unlit areas—these are where there is no street lighting or the lighting is less than Group B1 (typical B1 lighting—5 m columns spaced at approximately 33 m utilising lamp lumens of 3200–5300 lumens).

TABLE 1

A.P.L.E. RECOMMENDATIONS OF MAXIMUM LUMINANCE FOR ILLUMINATED ADVERTISEMENT SIGNS

Illuminated area (m^2)	*Zone* 1	*Zone* 2 (cd/m^2)	*Zone* 3 (cd/m^2)	*Zone* 4 (cd/m^2)
Up to 0·5		2 900	2 000	1 000
0·5 to 2·00		2 300	1 600	800
2·0 to 5·0	NO LIMIT	2 000	1 200	600
5·00 to 10·0		1 500	1 000	600
over 10·0		1 200	800	400

Note: 1 cd/m^2 = 0·291 9 *ft.L.*

It is important that the Local Authority Lighting Engineer has available a suitable instrument to check the luminance of signs to ensure compliance with the planning consent. The instrument should be capable of direct luminance measurements over a range from 10–10 000 cd/m^2 measured within $\frac{1}{2}$° or 1° field. It is a simple matter to reduce the luminance of a sign by installing tubular fluorescent lamps of a different classification. In general many signs are fitted initially with a White fluorescent lamp. The lumen output for lamp colours other than White may be obtained by applying the following multiplier to the relevant lumen output for the white lamps (65 W, 1·5 m White fluorescent lamp lighting design lumens—4700 lm).

		Colour rendering quality
Warm White	1·0	Fair
Daylight	0·96	Fair
Natural	0·75	Good
De-Luxe Warm White	0·68	Good
Kolorrite	0·51	Excellent
Softone 27; Warmtone	0·51	Excellent
Colour 32	0·6	Good
TL 84	1·02	Excellent

The installation of an alternative tubular fluorescent lamp to the White can reduce the luminance; but by the use of a lamp with good colour rendition the illuminated sign can be made more attractive, with greater sales potential.

A common error with applications for planning consent from petrol stations is that planning consent is given based on the illuminated

FIG. 9. Untidy layout.

FIG. 10. Good design: indirect lighting. MBFR lamps fitted within cylindrical supports on columns. (Courtesy Mobil Photo Library.)

advertisement signs. The project is completed and complaints of discomfort or disability glare are received by the Local Authority. The Lighting Engineer inspects the site and measures the luminance levels. The signs are quite satisfactory, but the area and overhead canopy lighting is the source of glare with bare fluorescent lamps. It is important that all lighting details are checked by the Lighting Engineer so that he can provide the necessary recommendations to the planner. If the street lighting is high pressure sodium he can recommend, say, mercury halide discharge lamps located within the canopy structure. The white light source will be more beneficial to the petrol station owner. Again if the street lighting has a white light source such as MBF/U lamps the canopy lighting can utilise 250 or 400 W high pressure sodium lamps. Figure 9 shows an example of bad layout and Fig. 10 is an illustration of a well designed petrol station with indirect lighting, the lamps concealed on support columns.

SITING LANTERNS

A worthwhile exercise in busy town and city centres is to install the street lighting lanterns on wall brackets fitted to the various building elevations (Figs. 11 and 12). This installation technique requires effort by the Lighting Engineer, but the end-product can be much more economic when

FIG. 11. Wall-mounted lanterns and service cupboards for street lighting. No columns.

FIG. 12. Wall-mounted street-lighting lantern.

compared with a conventional installation of lamp columns. It also means the street lighting system is made inconspicuous rather than cluttering up the footpath. The procedure required to obtain the consent of the property owner is contained within the Public Health Act, 1961, Section 45—Fittings for Street Lighting. The Section gives the owner who has consented to such works a right to require the fittings temporarily to be removed when necessary for reconstruction, repair or demolition of the building. The Local Authority must indemnify the property owner against any damage or loss arising as a result of the street lighting works. It obviously requires a small amount of 'salesmanship' by the Lighting Engineer to obtain the consent of the property owner. He can describe the inconvenience of 10 or 12 m lamp columns erected on the footpath and illustrate, with a photograph of a typical wall-mounted lantern, that the latter would be much less conspicuous. Multi-stores groups are the most difficult to convince, particularly with regard to the installation of the circuit cables. Again a photograph of a well-designed installation will enable the property owner to evaluate the quality of installation work. If consent is not given, the Local Authority will have the right contained within the Act to appeal to the magistrate against refusal of consent. It is very unusual to receive a refusal if a well-designed lantern, bracket and installation are proposed.

The cable installation for wall-mounted street lighting lanterns has to be carefully considered after surveying the various building structures. The first consideration will be the location for a single phase 240 V a.c. supply at 50 Hz. If a lamp column were to be erected the footpath would require opening up to obtain a service connection and this could cost in the region of £80 to £100 per electrical service. Add to this the charges for the lamp column erection and reinstatement and one can appreciate the extent of the

FIG. 13. Electrical service cupboard under window in passageway, for street lighting.

expenditure. The position of the existing electrical service within the properties should be recorded—many electrical services are located near to the front elevation on the ground floor and in older properties within the basement area. It is often possible to utilise the existing service positions. The Electricity Board can install an additional single phase fused cut-out direct to the unmetered side of the supply. A circuit cable can be installed from the point of supply to a service cupboard fitted flush into the external wall. An alternative external location may be a service cupboard concealed from normal line of sight within a passage or corridor of the property (Figs. 13 and 14). The circuit cables can be installed from the service cupboard and be connected to several wall bracket lighting points. The cable routes must be carefully designed to ensure the structure is not damaged and the installation is concealed as far as possible. Cable work with mineral insulated cables can be concealed within the cement joints of brick and stonework. If it is not possible to conceal the cables within the structure, the

FIG. 14. Typical purpose-built electrical service cupboard for street lighting.

cables can be contained within plastic ducting which can be tailored on the structure and painted to be inconspicuous. When the ducting is fitted into the vertical angles of the structure it is difficult to detect.

The control gear and photo-electric light cells can be concealed at high level in accessible positions for servicing by means of a tower wagon with a hydraulic controlled bucket. The siting position of the photo-electric light cell should not prevent its correct operation. An aluminium alloy cast box or glass fibre enclosure is suitable for the housing of the control gear. Care must be taken to ensure that mineral insulated cables (copper conductors and sheath) are not installed between the ignitor and the lamp terminals for high pressure sodium and mercury halide discharge lamps. The high ignition voltage would result in surge breakdown. PVC insulated and sheathed cables are recommended. A high pressure sodium lamp requires a peak voltage of from 1500 to 3000 V, depending on lamp type, to ensure ignition. The isolation of the wall bracket lantern from the control gear allows for greater freedom of design with the lantern. The wall bracket can be secured by means of non-ferrous hex head set screws with metal inserts fitted into the structure. On very old historic buildings the fixing bolts can be stainless steel passing through to the inside walls. Metal cross-straps can be fitted. The structure can be made good with additional structure supports if required. Great care must be taken to preserve historic

FIG. 15. Wall-mounted spherical lanterns for 400 W high pressure sodium lamps.

structures and the Department of the Environment may have to be consulted to approve the installation techniques. The design concept should produce a wall-mounted lantern with no untidy cables in sight (Fig. 15). The cable entry should be cut behind the bracket, and an architrave-type switch box is ideal for this purpose. If one considers that a 12 m tubular steel lamp column could have cost £130, the effort and job satisfaction of fitting wall bracket lanterns whenever possible is a worthwhile exercise.

Corrosion is a costly item and seldom given due consideration. It is therefore most important to design with materials which may be at a higher cost initially, but in the long term will be much more economical with low depreciation and longer life. Bracket arms can be aluminium alloy, galvanised or plastic-coated steel. Lanterns can be glass fibre, die cast aluminium alloy or high quality aluminium sheet, with all catches and hinges in stainless steel which does not corrode after a few years in service. Reflector systems should be in super purity aluminium. The lantern bowl can be in vandal resistant polycarbonate, and can be made watertight with silicon plastomer (anti-adherent material—age-resistant and resistant to ultra-violet radiation and preserving its properties to 250 °C). An active carbon filter can be fitted to prevent mineral particles etc. entering the lantern.

FIG. 16. Floodlighting of Castle Museum, Norwich. The structure is lit by close-offset 300 W PAR56 narrow-beam lamps, softened by SON/T (2 × 400 W) floods. Trees are lit by screened floodlights, halide and high pressure sodium.

FLOODLIGHTING

The floodlighting of buildings, trees and sculptures (Figs. 16 to 18) provides scope for the Lighting Engineer to demonstrate his artistic abilities, and the public at large will respond quickly and commend or condemn his efforts. The most frequent structure to be floodlit in a town or city is usually the cathedral or church building. The first important consideration is the annual charges for energy. There are numerous buildings in Great Britain where 40 kVA to well over 100 kVA of floodlighting has been installed. The energy charges are so high for church authorities that the number of hours when the floodlighting is in operation is often restricted to a few weeks per annum, and if they are situated in a tourist area the the tourist may never see

FIG. 17. Church floodlit by high pressure sodium lamps; monument by white light (sealed beam iodide).

FIG. 18. Riverside floodlighting. The beauty of reflections in water.

the floodlit spectacles. One should therefore give consideration to load factors which can be operated for 2000 hours per annum (dusk to midnight) at an economic level.

The light sources therefore should be mainly of the discharge types with high efficacy and long lamp life. This does not preclude the use of tungsten spotlamps etc. for the subtle modelling of features on a structure. At the present cost of electrical energy 1 kW of lighting in operation for 2000 hours per annum would cost £50. It may be that a church authority will pay a proportion of the energy charges, with the Local Authority being responsible for the remainder of the charges, although it can be said in general that most church authorities welcome floodlighting but do not accept the energy or maintenance charges. This differs from a bank or other commercial building where these charges are more readily accepted. A public building, such as a town or city hall, could much more easily absorb the charges for a modest floodlighting scheme. The existing maximum demand tariff structure would ensure an economical charge per kWh (approximately 2p per kWh). The load factor for floodlighting would be minimal if compared with the number of electric fires, kettles, etc., which can be found in oil and gas fired centrally heated buildings. The main economic light source choice could therefore be high pressure sodium (100 to 120 lm/W) or mercury halide lamps (67 to 95 lm/W).

The structure colour will often dictate which is the most suitable light source. The mercury halide with excellent colour rendition is very acceptable for flint-clad structures where the crisp white light highlights the numerous colours contained within the flints. The high pressure sodium has fair colour rendition and is more suitable for sandstone and similar coloured structures. The problem is to decide on a suitable economic light source and also to contrast the structure with its immediate surroundings. Thus if the street lighting is high pressure sodium and the structure floodlit with high pressure sodium the contrast will be reduced. If trees surround a church it may be possible to use mercury halide to floodlight the trees if high pressure sodium is utilised for the floodlighting of the church structure and the adjacent street lighting. It is therefore important to consider the colour rendition of the existing and possibly future street lighting prior to making a final decision on the light source to floodlight a building. A case in point is St. Paul's Cathedral which has high pressure sodium floodlighting adjacent to the high level, high pressure sodium street lighting in the City of London.

There are many manufacturers of floodlighting equipment in Great Britain and Western Europe. The quality varies considerably and it is futile to install any low cost equipment due to the operating conditions which the floodlight will have to endure in the British climate. It is advantageous if the front glass on the floodlight is fitted with a neoprene gasket and sealed to exclude water; the lamp holder etc. being accessible from the rear of the fitting. Thus the front glass and gasket are not disturbed to maintain the floodlight. The lantern should be locked with a special key—spring loaded clips etc. are the joy of vandals and lamps can disappear overnight. Most British floodlights are supplied in a basic bright metallic colour whilst a well-known German company supply all floodlights finished in matt black, a more suitable colour. A further omission from British manufacturers is the supply of masking louvres. These are a standard supply by many manufacturers in Western Europe. If floodlights are mounted at ground level they may be installed within a sunken enclosure or screened with a brick, stone, or wattle surround (Figs. 19a–b). It is important that the floodlight and control gear are securely bolted on to a concrete base as it is not beyond the power of vandals to remove floodlights. In a sunken enclosure such as a concrete pit there should be plastic conduits installed to enable adequate drainage away from the pit. When floodlights are fixed to structures above ground level the brackets, fixings and supports should be non-ferrous. A problem of fixing can be found in church and other roofs where the waterproofing material cannot be drilled. The floodlights must therefore be clamped to the structure, or hardwood timber supports fixed with an adhesive to the lead roof covering.

FIG. 19a. Concrete pit and hazel screen for floodlight.

When designing a floodlighting scheme it is necessary to select a level of illuminance. The brightness contrasts will be dependent upon the reflection factor of the structure and the location of the building in relation to its surroundings. Table 2 provides a guide for illumination levels.

There are two possible ways of calculating the types and number of floodlights required to achieve the desired illuminance; the lumen method and the luminous intensity method. For a large facade the lumen method should be used. This is based on a certain average luminous efficiency. For high and small objects—church steeples, sculptures, etc., the luminous

FIG. 19b. Purpose-built cobble enclosure for floodlight.

TABLE 2

Type of surface	*State*	*Illumination in lux*		
		Poorly lit surroundings	*Well lit surroundings*	*Brightly lit surroundings*
White marble	Fairly clean	25	50	100
Light concrete imitation	Fairly clean	50	100	200
White brick	Fairly clean	100	250	400
Yellow brick	Fairly clean	50	100	200
White granite	Fairly clean	150	300	600
Concrete or dark stone	Fairly clean	75	150	300
Red brick	Fairly clean	75	150	300
Concrete	Very dirty	Requires at least 150–300		
Red brick	Very dirty	Requires at least 150–300		

intensity method should be used. This is based on the luminous intensity radiated in a certain direction. The lumen method consists of calculating the number of lumens to be directed on to a facade in order to obtain a certain illuminance level. The number of lumens can be calculated by the formula:

$$\Phi_1 = \frac{F \times E}{\eta}$$

where Φ_1 is the total number of lamp lumens, i.e. the total luminous flux produced by all lamps, F is the surface of the facade to be illuminated in m^2, E is the desired illumination in lux on that facade and η is a factor which takes account of the efficiency of the fitting and the light losses (luminous efficiency).

The lumens produced by the lamps are concentrated by reflectors in the floodlights so that only 60 to 75 % of the initial lamp lumens are projected outwards and 40 to 25 % are lost through inter-reflection in the reflector system and absorption by other parts of the fitting. Finally a percentage of losses is accounted for by wasted light, that is light which is not incident to the building facade. The average utilisation factor in practice will be 0·25 to 0·35. By the use of the photometric data of the floodlights proposed, calculations can be more quickly and accurately tabulated by computer.

In the luminous intensity method the starting point is the luminous intensity, in candelas, radiated by a light source in a particular direction.

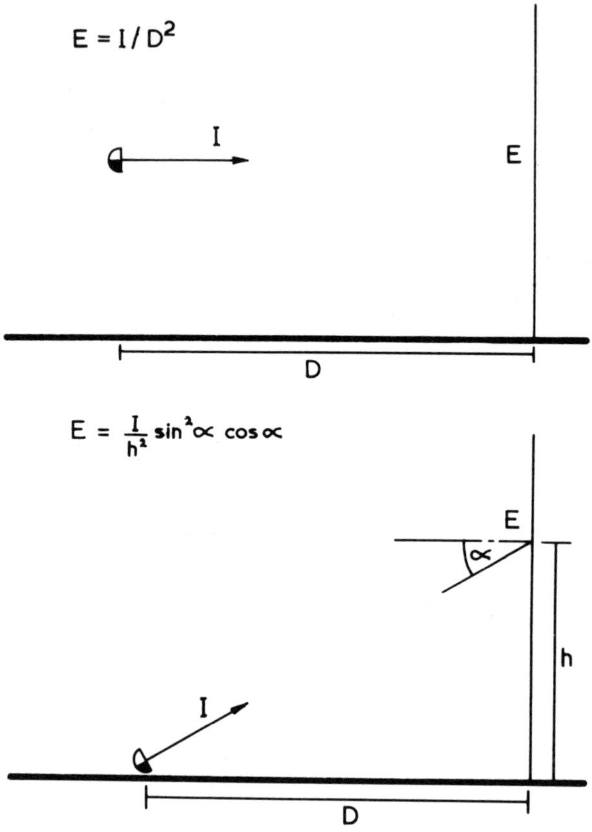

FIG. 20. The luminous intensity method.

The luminous intensity may be derived from a luminous intensity diagram or table. The calculation is made from the formula:

$$E = \frac{I}{h^2} \,.\, \sin^2 \alpha \,.\, \cos \alpha \qquad \text{and} \qquad \tan \alpha = \frac{h}{D}$$

where E is the vertical illuminance on the facade, I is the luminous intensity at the angle α, h is the height of the object above the level of which the fittings are arranged and α is the angle at which the light beam strikes the normal on the plane to be illuminated.

If, for example, the top of a church steeple has to be illuminated and the height of the steeple is 90 m above ground level, the floodlights could be placed on the roof of a nearby building say 20 m high, situated at 75 m from the tower. The required illuminance E is 50 lux. Given

$$h = 90 - 20 = 70 \text{ m}$$

$$\alpha = 43°,$$

then

$$E = \frac{I}{h^2} \sin^2 \alpha \cos \alpha$$

$$50 = \frac{I}{70^2} \times 0{\cdot}465 \times 0{\cdot}731$$

$$I = \frac{50 \times 4900}{0{\cdot}34}$$

$$= 720\,000 \text{ candela.}$$

The luminous intensity in the beam centre should be 720 000 cd or two floodlights each giving half this figure. Floodlights can be obtained with various light distributions—such as wide, medium, narrow and elliptical angle beam distributions. There are also tungsten and tungsten halogen and compact source iodide sealed-beam lamps with wide, medium and narrow beam distributions. Thus we have the lamps and equipment, with low energy consumption, to provide spectacular scenes for pleasure, advertising and security.

Britain's earnings from tourism are one obvious reason why every effort should be made to preserve and improve the lighting systems which do so much to enhance the environment. Effective measures against vandalism are vital to prevent the wilful destruction of costly lighting systems and other civic amenities. The full co-operation of planners, architects and lighting engineers is necessary if the day-time and night-time environments are to be protected and improved.

ACKNOWLEDGEMENT

The permission of the Civic Trust to print an extract from their *News Sheet* is gratefully acknowledged.

REFERENCE

1. *Civic Trust News Sheet*, September, 1972.

Chapter 7

THE ENGINEERING OF DAYLIGHT

J. LONGMORE
F.C.I.B.S., A.R.P.S., F.R.S.A.

Senior Lecturer in Environmental Design and Engineering, Bartlett School of Architecture and Planning, University College, London, UK

SUMMARY

The general principles of daylighting in buildings are discussed and consideration is given to recent developments in daylighting design, the integration of artificial light with daylight, the thermal aspects of daylight design, and energy conservation. The well known daylighting prediction techniques are briefly referred to and emphasis is given to the need for simple procedures for window sizing in the early stages of design rather than for precise methods of testing a proposed solution. The concepts of average daylight factor and daylight diversity within a room are discussed in some detail and the compatibility of design procedures for daylighting and artificial lighting is considered essential in relation to the majority of present-day buildings in which the two sources of light are used together for much of the time. In this context the lumen method of daylighting design (or daylight flux admittance method) is recommended as a means of generating window design in the early stages of the architectural design process, which is non-coercive and does not dictate a single 'right' solution.

Recent consideration given to the consumption of energy in buildings and the need for energy conservation has created circumstances which give rise to a revitalised interest in windows and daylighting design—on the one hand to make more efficient and effective use of the daylight which is available to buildings and thus to reduce the consumption of electricity for artificial lighting, and on the other hand to develop means of controlling and selectively utilising the solar heat energy admitted by windows so that it may serve to reduce fuel consumption for heating during the winter months without creating excessive solar gain problems in summer. Much of the necessary data already exists and many of the basic principles are well known and their successful application merely calls for care in design and for a reversal of the trend towards designed dependence on permanent artificial lighting and air-conditioning in many building types where the latter is not an essential requirement for climatic reasons.

The effective use of daylight can reduce the consumption of energy in buildings and it is notable that the Illuminating Engineering Society in the United States is currently revising its 'Recommended Practice of Daylighting' to take account of the need for greater reliance to be placed on daylighting in buildings as an energy conservation measure.

Daylight differs fundamentally from artificial light as a source of lighting in buildings in that it originates from a source external to the building and some provision must be made for its admission to the interior. For this reason the engineering of daylight is essentially interrelated with the engineering of the building itself and decisions on daylighting design have a considerable influence on the building design.

In all consideration of daylighting design, therefore, it is necessary to relate the exterior to the interior of the building, and to recognise that daylighting in buildings comes from an area source of light (the sky) which is external to the building and which requires some aperture (the windows) to be provided in the building envelope in order to admit light to the interior. In general terms, the larger the aperture the more light is admitted and the window size, its location and the transmission characteristics of the glazing material are the principal factors influencing daylight quantity which are under the control of the designer. Other factors, however, such as external obstruction have a significant effect on the amount of light available at the window and must also be taken into account.

The ultimate source of all daylight at any latitude is the sun, and daylight (or sky light) is the result of differing degrees of scattering of the sun's rays in the atmosphere. In the hot, arid, tropic regions direct sunlight forms the greater part of light reaching the earth's surface and only a small proportion is scattered by dust with the result that the sky is deep blue, of low luminance and lighting is harsh and gives deep contrasts.

In more humid areas an increasing proportion of the sun's rays is scattered by dust and moisture in the atmosphere until the condition is reached where they are totally diffused by layers of cloud to give the densely overcast sky common in temperate regions and in the humid tropics. In temperate regions, such as North-West Europe, the densely overcast sky has become accepted as the standard source of daylight for design purposes because it represents the worst lighting conditions, but there is a need to give greater attention to the improved conditions provided by other sky conditions particularly in terms of reducing the need for artificial lighting.

The illuminance provided by the sky is not constant, as is that from electric lamps, but varies continuously and this variation is one of the fundamental differences between daylighting design and design for artificial lighting. The luminance of the sky, and hence the illuminance which it provides, varies with the height of the sun above the horizon and so depends on latitude, on time of day and on the seasons. In addition to these systematic and predictable variations for the overcast sky there are also random variations due to the density and movement of cloud cover at any given time.

The illuminance on a horizontal surface from the whole sky out-of-doors in the London area reaches a maximum of about 60 000 lux in summer (the monthly average at noon in July is about 35 000 lux) but rarely exceeds 20 000 lux in winter (the monthly average at noon in December is about 9 000 lux).[1] The duration of 'useful' interior daylighting, i.e. when the outdoor illuminance exceeds 5000 lux, is about 14 h per day in mid-summer and drops to less than 5 h per day in mid-winter so daylight in buildings must necessarily be supplemented by artificial light for part of the day for at least part of the year.

The illuminance at points in a room lit by windows in side walls is rarely more than one-tenth part of that out-of-doors and is often considerably less at points far from the window; nevertheless it is of significant magnitude for much of the year.

DAYLIGHT FACTOR

Because of the variations in illuminance from the sky it is customary to express interior daylighting in relative rather than absolute terms, as a fraction—usually a percentage—of the illuminance occurring simultaneously on a horizontal surface outdoors exposed to the whole sky. This fraction is known as a 'daylight factor'; it is a geometrical property of the room and of the window and, for a given sky luminance distribution such as that of the standard overcast sky, it is effectively a constant at any particular point within the room and is independent of changes in the sky illuminance.

(a) the sky component

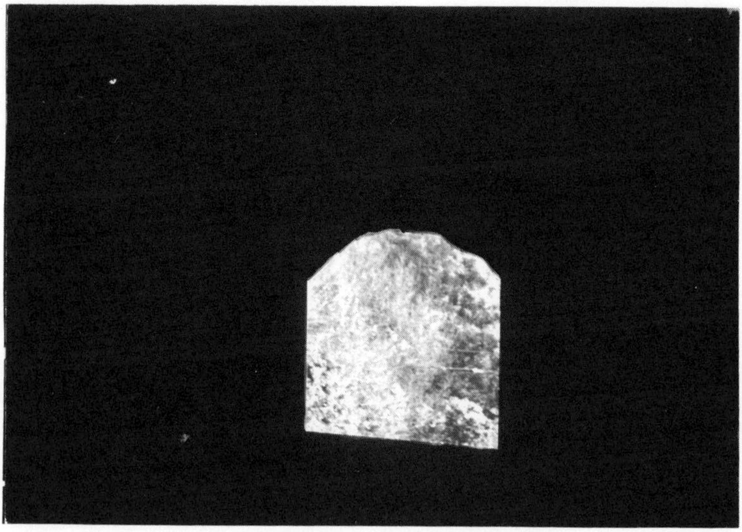

(b) the externally reflected component

FIG. 1. The constituents of the daylight factor.

(c) the internally reflected component

(d) the total daylight factor is the sum of the three components

FIG. 1—*cont.*

The daylight factor is the ratio of the daylight illuminance E_i at a point on a given plane due to the light received directly or indirectly from a sky of assumed or known luminance distribution, to the illuminance E_o on a horizontal plane outdoors due to an unobstructed hemisphere of this sky, i.e.

$$\text{daylight factor} = \frac{E_i}{E_o} \times 100\,\%$$

The interior illuminance at a point in a room can thus be derived from the daylight factor (DF) for any given outdoor illuminance or for any particular time of day and month of the year—from measured sky illuminance data, i.e.

$$E_i = \frac{\text{DF} \times E_o}{100}\ \text{lux}$$

The daylight factor at any point in a room depends on the area (or angular dimensions) of the patch of sky which is visible through the window from that point and generally comprises three components (see Fig. 1):

1. The sky component, or light reaching the point directly from the sky.
2. The externally reflected component, or light reaching the point after reflection from external surfaces which are obscuring part of the sky that would otherwise be visible through the window.
3. The internally reflected component, i.e. light admitted by the window which reaches the reference point after reflection from surfaces within the room.

The relative magnitudes of the three components depend primarily on the location within the room of the reference point, the sky component being greatest near the window and least near the back of the room, whereas the internally reflected component is almost uniform throughout the whole room and depends on the reflectances of the interior surfaces. The externally reflected component is usually small and depends on both the degree of external obstruction and its reflectance, or its luminance relative to that of the sky which is obscured.

Thus interior daylight varies at different points within a room not only in quantity but also in quality, being most highly directional and creating a more pronounced degree of modelling near the window, whereas near the back of the room the light is softer and more uniformly diffused from all directions. Because of the directional character of daylight and because of the diversity between points near to a window and those at the rear of a room, reflected light is very important, much more so than in artificial lighting, and in daylighting design the reflectances of the principal room surfaces form an integral part of the design.

Reflected light also helps to reduce harsh modelling and contrasts around the window which give rise to glare for directions of view towards the window. It also reduces the diversity of daylight factors from one side of a room to another and, at points far from the window, it may represent the largest single component of the daylight factor.

DAYLIGHTING DESIGN

Developments in daylighting design have tended to be based on attempts to achieve greater engineering precision in the determination, or prediction, of the three components of the daylight factor at a point. Techniques such as the Waldram Diagram, the BRS Daylight Protractors, the BRS Split Flux Method and the Pilkington Dot Diagrams have all been developed for this purpose. However, they are not so much aids to design as simply testing procedures to be applied after the design decisions have been made, possibly on an intuitive basis or on some criteria other than daylighting.

Daylighting and windows can no longer be considered as isolated elements of a building design and it is necessary to be able to examine at an early stage in design the implications and interactions of daylighting design decisions with other design criteria such as artificial lighting, energy consumption, heat loss, heat gains, sound transmission and economics. The need to balance daylighting and visual requirements with some of the consequential disadvantages of glass in the building envelope is much greater now than previously. Considerable emphasis has always been placed on the provision of daylight in buildings, and in those situations where daylight is inadequate as a source of interior illuminance it has been supplemented by artificial light rather than by adopting the perhaps more attractive engineering solution of designing buildings entirely without windows.

There is now a generally closer integration of artificial lighting with daylighting design in many building types and it is significant that the Code for Interior Lighting of the Illuminating Engineering Society[2] and the IES Lighting Guides take more account of daylighting now than similar publications have done for many years. The availability of fluorescent tubular lamps, producing light suitable in spectral quality and in quantity for daytime use, has diminished the emphasis on daylight quantity and has enabled greater attention to be given to the qualitative aspects of daylighting design such as freedom from glare discomfort and disability, view out, directional characteristics of light within interiors, surface luminances and temporal and spatial variations.

It seems likely that increasing emphasis on energy conservation will lead to greater reliance being placed on daylight as a naturally occurring renewable form of solar energy and to a re-appraisal of deep-plan artificially lit and air-conditioned buildings which are highly dependent on the use of electricity.

THE INTEGRATION OF ARTIFICIAL LIGHT WITH DAYLIGHT

It has long been recognised that daylight admitted by windows in one side wall of a room cannot provide at all times and at all points in the interior the comparatively high illuminances which are currently recommended, particularly for work places. Rising standards of task illuminance, occurring at a time when deep-plan buildings with low ceilings were being developed, led to a situation in which it was necessary for artificial lighting to be used in such buildings throughout the entire working day as the primary source of task illuminance, and to a consequent reduction in window areas.

In most buildings in which, for various reasons, daylight is inadequate as a source of working illumination, the designer should aim to use daylighting and artificial lighting together. The daylight and artificial light serve different functions but they can be compatible and complementary and each can and should be used to compensate for the deficiencies of the other. The IES Technical Report No. 4[3] sets out some basic data and design procedures for the integration of artificial light with daylight in a form designed to assist all those concerned with the lighting of buildings to appreciate the daytime aspects of the subject. It provides sufficient detail to enable design decisions to be taken without reference to other publications.

In general, whereas in the past the attainment of a specified minimum daylight factor, usually at some point farthest from the window, has been considered essential to daylighting design in most building types, this is no longer the case and the ready availability of artificial lighting enables daylight to be supplemented in those areas most difficult to light satisfactorily from side windows. The result is that the minimum daylight factor need not be a critical design consideration other than in interiors which are specifically required to be lit by daylight alone.

The integration of artificial light with daylight for continuous use during daylight hours has been made possible largely because of the availability of types of fluorescent tube whose colour characteristics are closely similar to those of daylight. However, some of the more recently developed types of compact discharge lamp are less satisfactory in this respect—though improvements are probable—and their colour characteristics combined with their small size and high intensity necessitate some considerable care in their use if visual integration with daylight within the interior is to be achieved.

DAYLIGHT AND ENERGY CONSERVATION

Daylight may be regarded as a freely available, naturally occurring, renewable form of solar energy but its utilisation is not without disadvantages. Daylighting and windows are often criticised because of the loss of heat energy which results from the poor thermal insulation value of single glazing, but in many building types, and particularly in housing, daylight alone is quite adequate for the interior lighting requirements during normal daylight hours and any significant reduction in window sizes would inevitably lead to an increase in the use of artificial lighting throughout the year.

The deep-plan Integrated Environmental Design concept of building design is highly energy dependent and is based on excess heat production, by the use of electric lighting within a building throughout much of the year, with a consequent need for cooling even in winter. Energy conservation and optimal environmental design considerations indicate that the Integrated Environmental Design principles should be carefully re-examined before being extended to other building types in a climate which in itself creates little or no need for artificial cooling in buildings with low energy consumption.

Windows are an integral part of the fabric of a building and must be considered from the initial planning stage, as decisions on daylighting and windows can profoundly affect the plan form; servicing and economics and any faults or inadequacies in the daylighting design can rarely be rectified after completion of the building.

It would appear to be opportune to give serious consideration to some of the applications of daylighting in buildings which have tended to fall into disuse as a result of the ready availability of electric light and the ease with which it can be added—without structural alteration—at a comparatively late stage in design.

For example, the practice of using 'borrowed' light to provide background lighting in internal spaces and in corridors in particular is worthy of reappraisal (see Fig. 2). Glazed panels above doors and in the upper parts of the internal walls of cellular offices and similar spaces serve to transmit daylight, and also some borrowed artificial light, to spaces which are otherwise wholly dependent on permanent artificial lighting and which perhaps necessitate emergency lighting. Windows at the ends of internal

FIG. 2. Borrowed light in a corridor. (Courtesy Pilkington Brothers Limited.)

corridors and roof lights over internal stairways provide a useful level of safety lighting in those areas where some form of emergency lighting provision is essential. It is perhaps not inappropriate to suggest that the requirement for staircases in dwellings to be placed against external walls with windows might be extended to other building types.

The directional character of daylight from side windows can lead to an under-evaluation of the quantity and quality of daylight in a room, due to the cosine effect, if measurements or calculations are referred only to a horizontal plane. However, such uni-directional lighting is of considerable significance in relation to the elimination of specular reflections in the visual task. The elimination of veiling glare can lead to higher visual effectiveness of the lighting such that, in situations where reflections can be largely avoided by side lighting at right angles to the direction of view, lower illuminances may be satisfactory than those recommended for general artificial lighting from the ceiling.

Energy conservation makes it all the more important to ensure that the light transmission of windows and of roof lights in particular should be maintained as high as possible by regular cleaning, and suitable access for cleaning must be incorporated into the design.

THE DISTRIBUTION OF DAYLIGHT IN A ROOM

Daylight in a room lit by windows in one side wall varies in a systematic fashion with distance from the window wall. This distribution or spatial variation can be conveniently expressed numerically as a daylight diversity or uniformity ratio following standard practice in artificial lighting design, and it may be defined in several ways, for example as the ratio of maximum to minimum daylight factor, of maximum to average daylight factor or of minimum to average daylight factor. Experience has demonstrated that minimum and average daylight factors have greater significance in design and visual appreciation of an interior than the maximum, and thus the most useful expression for daylight diversity would appear to be the ratio of the minimum to the average daylight factor.

The diversity of the daylight in a room depends on a number of factors, of which the most important are the shape of the room, the positions and sizes of windows or roof glazing, the reflectances of the interior surfaces and the luminance distribution of the sky. The aim in daylighting design should be to manipulate these various factors in such a way that both the average daylight factor and the diversity throughout the room are appropriate to the activity which it will accommodate and the extent to which the daylight is to be supplemented by artificial light.

DAYLIGHTING PREDICTION TECHNIQUES

Daylighting prediction techniques have become increasingly sophisticated over the years and enable daylight factors within spaces to be determined with a high degree of precision despite the complex nature of the problem. This very sophistication, however, has worked against their practical application in design and has led to certain basic simplifying assumptions being made whereby the predicted daylight factors apply to strictly limited sets of conditions which may not necessarily be adequately representative of the wide range of conditions that can prevail. Calculations are usually based on an assumed CIE standard overcast sky luminance distribution and are generally restricted to a horizontal reference plane—often to a limited part only of the reference plane remote from the windows—primarily as a test of compliance with a specified minimum daylight factor. Such calculations take little or no account of the distribution of daylight in the room or of the suitability of the room depth to the required standard of daylighting, but most importantly they tend to neglect completely the areas of the room nearer to the window where the daylighting may well be excessive and within which the large glass areas most probably give rise to visual and thermal discomfort. For many interiors it is necessary to supplement these stylised calculations by constructing a scale model in which the visual qualities of the daylighting in the space as a whole can be examined, if possible, under a variety of sky conditions.

Certain simplified prediction techniques exist, such as the BRS sky component tables[4] and the internally reflected component tables or nomograms, but none of these is so universally applicable in the preliminary stages of design as the lumen method of artificial lighting design. Lumen methods of daylighting design have been proposed many times in the past and have been used in the USA for a number of years to determine maximum, minimum and average daylight illuminances in interiors at various times of the year and for different sky conditions.[5]

Lumen methods of daylighting design have not found favour elsewhere, however, because it has been assumed that spatial diversity and temporal variations render average illuminances or average daylight factors of limited value. But it can equally be argued that the minimum daylight factor based on overcast sky conditions is also of limited value and the basis of such objections to lumen methods of daylighting design has been undermined, if not entirely removed, by the possibility of integrating artificial light with daylight. Several important benefits can be shown to accrue from a common approach to the design of daylighting and artificial lighting, at least in the initial design stages, for those interiors in which daylight is not intended to be the sole source of illumination. For most interiors the ready availability of artificial lighting now means that the point of minimum daylight factor farthest from the window is, in daylighting terms, the least important rather than the most important point in the room.

The Commission Internationale de L'Eclairage published in 1970 'International Recommendations for the Calculation of Natural

Daylight'.[6] The aim of these recommendations was to reduce a few of the available techniques for predicting daylight into a method simple enough to apply to everyday lighting design problems.

For interiors with roof lighting the method was based on material previously published in Australia.[7] It gives the average daylight factor on a horizontal plane and is based on a uniformity ratio (maximum: minimum) of the order of 2:1.

AVERAGE DAYLIGHT FACTOR

The visual impression of a room lit by daylight from side windows is governed to a great extent by the brightness (relative luminance) of all the surfaces in the room in relation to the prevailing brightness conditions outdoors as seen through the windows. This visual integration of interior brightness is related more closely to the average daylight factor in the room than, for example, to the maximum or more particularly to the minimum daylight factor. The average daylight factor can therefore be used as a means of assessing or predicting the visual adequacy of the daylighting in the room as a whole rather than at any particular, but not necessarily representative, point.

The use of minimum daylight factor alone as a criterion of the adequacy of interior daylighting has developed as a matter of computational convenience but it has serious limitations because it gives no indication of the adequacy of the daylighting in the room as a whole, without a full knowledge of the fenestration in relation to the shape and size of the room and of the interior surface reflectances. A marked improvement in the specification of interior daylighting, over the minimum daylight factor alone, is achieved if two figures are used—the average daylight factor and the uniformity ratio (minimum: average). There is an added advantage in that this form of specification corresponds to artificial lighting practice and so enables comparisons to be made which are of particular value in interiors which are to be lit partly by daylight and partly by permanent artificial lighting.

The average daylight factor can be defined in a number of ways but it is perhaps convenient if it is determined to a first order of approximation at the design stage as:

$$\frac{\text{total incident light flux on reference plane}}{\text{area of reference plane} \times \text{outdoor illuminance}} \times 100\%$$

The diversity of the daylighting on any plane in a room is a function of the room depth measured from the window wall, the interior surface reflectances and also of the luminance distribution of the sky. The standard overcast sky has a zenith luminance which is three times greater than that at the horizon, so a diversity (uniformity ratio) based on the overcast sky will be materially changed, especially in a side lit room, if bright sunlit clouds are visible in the zone of sky near the horizon. The standard overcast sky is generally assumed to produce the worst daylighting conditions which are likely to be experienced in a side lit room and consequently under most other sky conditions the diversity of daylight will be improved, i.e. the daylight factor distribution will be more uniform. Any change in the average daylight factor, however, is likely to be quite small.

The visual appearance of a room, and the judgement as to whether it is satisfactorily lit depend on both the average daylight factor in the room and the diversity of the lighting (uniformity ratio) and are influenced by the reflectances and distribution of light on the interior surfaces. If the average daylight factor is appropriate to the function of the room but the uniformity ratio is low, it will appear badly lit and gloomy at points far from the window. Visual satisfaction can be improved, for a given average daylight factor, if the uniformity ratio is as high as is practicable.

If a requisite average daylight factor is achieved by the daylighting design but the uniformity ratio does not meet the recommended value, permanent artificial lighting will be needed as a supplement to the daylight in the areas remote from the windows and will be more essential on bright days than on dull days. On the other hand, if the daylighting design achieves the required uniformity ratio but the average daylight factor is less than that recommended, the artificial lighting supplement will be necessary over a greater part of the room but for limited periods only when the outdoor illuminance is low. A distinction can therefore be made between artificial lighting which is used to supplement daylight spatially and that which is used to supplement it temporarily.

The average daylight factor in a room lit by windows in a side wall can be found, to a first order of approximation, from an adaptation of the BRS 'split flux formula':

$$\text{average daylight factor} = 0{\cdot}85W\left(\frac{C}{A_{\text{fw}}} + \frac{CR_{\text{fw}} + 5R_{\text{cw}}}{A(1 - R)}\right)\%$$

where W is the actual glass area in the window, A is the total area of all the interior surfaces, A_{fw} is the area of the floor and those parts of the walls below the mid-height of the window (not including the window wall), C is a function of the flux incident on the window plane from above the horizontal and varies with the sky luminance distribution and the external obstruction angle, R_{fw} is the average reflectance of the floor and the lower walls (not including the window wall), and R_{cw} is the average reflectance of the ceiling and upper walls (not including the window wall).

The values of the coefficient C for a CIE standard overcast sky are given in Table 1.

TABLE 1

Angle of obstruction as seen from centre-line of window (degrees above horizontal)	*Coefficient* C
0° (no obstruction)	39
10°	35
20°	31
30°	25
40°	20
50°	14
60°	10
70°	7
80°	5

Alternatively, the calculation can be carried out to determine the average daylight illuminance by multiplying by an appropriate factor according to the particular outdoor sky illuminance. For a standard overcast sky of 5000 lux outdoor illuminance the formula becomes:

$$0{\cdot}85 \times 50W\left(\frac{C}{A_{fw}} + \frac{CR_{fw} + 5R_{cw}}{A(1 - R)}\right)\text{lux}$$

COMPATIBILITY OF DESIGN PROCEDURES

Artificial lighting is used continuously throughout the day in many buildings to provide task illuminances or to relieve unsatisfactory daylight diversity and harsh modelling from the windows, or both. It is highly desirable therefore that the design procedures for daylight and for artificial light should be compatible and that daylighting should be capable of being expressed in terms which are familiar to the lighting engineer and which may lead to a better understanding of daylighting design by lighting engineers.

Artificial lighting design is generally based on a standard service illuminance, averaged over the area of the reference surface, and on a recommended uniformity ratio, expressed as minimum:average illuminance, which for general lighting installations should not be less than 0·8.

There is a need to examine carefully some of the assumptions which are made about uniformity, both in artificial lighting design and in daylighting design, and particularly in interiors lit by combined daylight and artificial light. For example in an interior designed to be lit by daylight from side windows and artificial light from the ceiling, the two sources of light necessarily have different distribution patterns which to a certain extent are complementary—the daylight serving to buffer the diversity of artificial light and the artificial light buffering the diversity of daylight from one side of the room to the other. The effective uniformity ratios for the combined installation are therefore better than those for either source alone, offering greater freedom of design of both the windows and the artificial lighting installation.

Recommendations for permanent artificial lighting in daylit interiors might well be based, therefore, on a lower uniformity ratio than the customary value of 0·8.

Average daylight factor and daylight uniformity ratio are concepts which have been generally considered to be of small significance for daylighting design in rooms lit by windows in one side wall because of the considerable diversity across the room expressed as maximum:minimum. Daylighting design has therefore developed technologies based on point-by-point calculation techniques for the determination of daylight distribution and particularly of minimum values. In practice, however, the design has been largely concerned with testing compliance with recommended minima and has been concentrated primarily on the worst lit point farthest from the window, taking little or no account of distribution, of the maximum values occurring near to the windows or of the glass areas necessary to achieve the minimum standard in rooms of practicable depth. This practice can lead to over-fenestration, to unsatisfactory thermal performance of the building and to a daylight diversity which makes artificial lighting necessary even in rooms which were designed to be adequately lit by daylight, because of the adverse adaptation effects generated by the areas of maximum illuminance near to the windows and to the glare resulting from the considerable views of sky afforded by large glass areas.

Average daylight factor and daylight uniformity ratio can be shown to be valid concepts in daylighting design, for both the architect and the lighting engineer, and particularly in the context of greater use of artificial light within buildings during daylight hours. Average daylight factor bears a closer relationship to visual impressions of quantity and quality of light within a room than does the daylight factor at any single point and, when coupled with a limiting uniformity ratio, can be used to define the distribution of light throughout the greater part of the space with the possible exception of those areas in close proximity to the windows.

However, it is necessary to consider also the distance over which the uniformity ratio applies. In a room lit from windows in one side wall only, the distance between points at which the daylight factor corresponds to the average value and those where the daylight factor is a minimum will be at least half the depth of the room. If the same room is lit by a conventional ceiling-mounted artificial lighting installation, the respective distance between points of average and minimum illuminance is considerably smaller. By considering the diversity per unit distance, or 'diversity gradient', it is possible to compare diversity in daylit and artificially lit interiors on a more equitable basis. The diversity gradient for side windows applies to the whole room whereas that for roof lit or artificially lit interiors applies in most instances merely to the distance between adjacent roof lights or luminaires or to their distance from the side walls. The real diversity gradient for artificial lighting or roof lighting can thus be significantly greater, and visually more apparent, than that for side windows unless it is masked by changes in surface reflectances or other interior detailing. The greater the distance over which the diversity applies, the greater the diversity may be, within certain overall limits.

A convenient method of assessing the daylight quantity in a room which enables comparisons to be made with the design of a permanent supplementary artificial lighting installation, particularly during the preliminary stages of design, is one which expresses the daylighting in terms of the 'daylight flux admittance' of the window, i.e. a lumen method of daylighting design.

Consider, for example, a room 8 m × 6 m × 3 m high, lit from a window in one long wall of which the glass area is 25 % of the external wall area, i.e. 6 m^2. The window is of clear glass, the diffuse light transmittance of which is 0·85. The incident illuminance on the window plane on a dull overcast day in winter is approximately 2200 lux when there are no external obstructions. The total daylight flux admitted to the room by the window is therefore 2200 × 6 × 0·85 = 11 000 lm (approx).

In summer, on a bright cloudy day, the incident illuminance on the window may be of the order of six times that on a dull day in winter, so that the daylight flux admitted to the room would be about 67 000 lm.

To simplify comparison with the artificial lighting, it may be assumed that the room is to be lit to an average illuminance of 400 lux by an artificial lighting installation using luminaires whose light output ratio (LOR) is 0·7 and utilisation factor (F_u) is 0·5 for room surface reflectances of the ceiling, walls and floor of 70 %, 50 % and 30 % respectively.

The light flux emitted into the room by the artificial lighting installation is given by:

$$\frac{E_h \times A \times \mathrm{LOR}}{F_u \times F_m} = \frac{400 \times 48 \times 0{\cdot}7}{0{\cdot}5 \times 0{\cdot}8} = 34\,000\ \mathrm{lm}$$

where F_m is the maintenance factor.

Thus it can be seen that the amount of light admitted to the room by the window is less than the flux emitted by the artificial lighting installation when the outdoor sky illuminance drops below about 15 000 lux, i.e. for 47 % of normal working hours between 09.00 and 17.30. The room as a whole is therefore likely to appear to be dominantly lit by the artificial lighting installation for about one half of the time and dominantly lit by daylight for the remainder. The basic difference between the effects of the two sources lies in the way that the flux is distributed throughout the room and in the relative angles of incidence on the principal room surfaces. The average daylight factor for the room is found from the modified split flux formula to be 3·3 %, producing an average daylight illuminance of 165 lux on a dull overcast day, rising to about 1000 lux on bright, cloudy days in summer.

The light output ratio of a window may be regarded as equivalent to the diffuse transmittance of the glass, i.e. 0·85 for clear glass and some lower value if allowance is made for the window frame and glazing bars within the window aperture. The lower flux fraction of a window depends on the angle of elevation of external obstructions and may be taken as $(0{\cdot}39 \times 0{\cdot}85)/(0{\cdot}44 \times 0{\cdot}85)$, i.e. 0·886 for an unobstructed window glazed with clear glass, the upper flux fraction would be $(0{\cdot}05 \times 0{\cdot}85)/(0{\cdot}44 \times 0{\cdot}85)$, i.e. 0·114. For cast or diffusing glass the upper and lower flux fractions would be more nearly equal.

The flux admittance of a window is a function of the window and not of the room in which it is used, being dependent only on the sky luminance distribution, the daylight illuminance on the outer surface of the window, the glass area, the type of glass and state of cleanliness, and any external obstructions and projections. It is independent of the shape of the window and of the shape and size of the room. Thus it is possible to describe windows in terms of their light admitting potential, e.g. as a 10 000 lumen window, or as 1870 lm/m^2 of glass area for an outdoor horizontal illuminance of 5000 lux from the whole sky, and to generate the window areas appropriate to the average daylighting requirements of a room during the early stages of design.

Daylighting design at the sketch design stage can be usefully based on the flux admittance method and on window potential in terms of average daylight factor, i.e. that a window of given area is capable of providing an average daylight factor of $d\%$ within an area of $a\,m^2$ if assumptions are made about surface reflectances. The designer can thus synthesise his design from first principles in a sequential and systematic fashion rather than by the trial and error procedures employed in testing compliance with minimum daylight factor standards.

Lynes[8] has proposed a simplified lumen method of daylight design for side lit rooms which is based upon an essentially similar approach to that described above. He expresses the illuminance on the outside surface of the window as:

$$\frac{S\,.\,E_o}{100}\,\text{lux}$$

where S is the daylight factor on the outside surface of the window (per cent) and E_o is the outdoor illuminance on a horizontal surface from the unobstructed overcast sky.

If the luminance of the ground and outdoor obstructions is about one-tenth of the average luminance of the overcast sky, S is given approximately by the expression:

$$S = \frac{\theta}{2}\,\%$$

where θ is the vertical angle, in degrees, subtended at the centre of the window by the directly visible sky.

The illuminance on the outside surface of the window then becomes $(\theta\,.\,E_o)/200$ lux and the daylight flux entering the window is:

$$\frac{W\,.\,t\,.\,\theta\,.\,E_o}{200}\,\text{lumens}$$

where W = window glass area (m^2) and t = transmittance of window glass.

The significance of a lumen or flux admittance method of daylight design is that it makes possible a three-stage procedure:

1. To determine θ, the vertical angle subtended at the centre of the window by the visible patch of sky and to express the daylight factor on the outside surface of the window in terms of this angle. The value of θ is influenced by the siting and orientation of the building in relation to its surroundings and will be reduced by any horizontal limitation on the patch of visible sky, but offers a means of comparing alternative building forms from the point of view of daylighting requirements.
2. The second stage in the daylight design is to estimate W, the window area which will depend upon the chosen average daylight factor. At this stage the total glass area is in question, not the shape or disposition of the windows. Many other factors will interact to control the window area, notably thermal considerations such as heat loss and solar gain and the optimum must be a compromise. Double glazing or heat-rejecting glass will involve a reduction in transmittance and a corresponding increase in the window area for a given flux admittance but the determination of window area at this stage ensures that daylighting considerations will be given due weight in relation to thermal and other factors. If the window area is judged to be excessive for other reasons the consequence can be faced that the daylighting design targets cannot be met and either the plan form of the building must be changed or permanent supplementary artificial lighting must be used.
3. The third stage in the daylight design is to decide the shape and position of the windows whose area has already been determined. If daylighting considerations are paramount, attention must be concentrated upon achieving a good uniformity of daylight distribution.

By comparison with traditional methods of daylight prediction the daylight flux admittance or lumen method of design has three clear advantages:

1. It does not share the disadvantages of the Waldram diagram, BRS daylight protractors or overlay techniques, of delaying daylight design calculations until the shape, size and position of the windows have already been determined and can be shown in plan, section or elevation. It is capable of generating window design and is not restricted to the more passive role of checking for compliance with a code of practice.

2. It enables daylighting considerations to be incorporated at each stage of the architectural design process, from the site analysis to the interior design.
3. It is non-coercive and does not dictate a single 'right' solution, but encourages a designer to weigh the conflicting requirements in a logical sequence.

Lynes[9] has suggested a method of determination of uniformity of daylighting in a room which obviates the need to make at least one point-by-point calculation as is required for example by the uniformity ratio of minimum:average daylight factor. His proposal is that uniformity should be judged in terms of the ratio of the average daylight factor in the half of the room which includes the window to the average daylight factor in the other half of the room. This method is applicable to rooms lit from windows in one side wall only but not to more complex window arrangements.

THERMAL ASPECTS OF DAYLIGHT DESIGN

Energy conservation considerations have undoubtedly revitalised interest in daylighting design and in architectural and engineering solutions designed to maximise the utilisation of the daylight which is available to and within buildings. This in turn will lead to a closer and more systematically controlled relationship between daylighting and artificial lighting by ensuring, for example, that artificial lighting can be switched off in areas and at times when it is not required and that unwanted solar heat gains through windows can be selectively controlled.

Wilberforce[10] has examined the energy balance of glazing by comparing the radiative heat gains with the conductive heat losses and has demonstrated that the radiative gains can make a significant contribution to energy requirements within buildings for certain times of the year.

Glazing admits energy to buildings by the direct transmission of solar radiation and allows energy to be lost by conduction when the outdoor temperature is lower than that within the building. These two processes occur simultaneously but the problem in predicting whether the net exchange of energy is a gain or a loss, has been that until very recently there were no data available of solar radiation incident on vertical surfaces in the UK. The data which are now available are of direct plus sky diffuse radiation incident on horizontal and on north, east, south and west-facing surfaces at the Meteorological Office, Bracknell.

These data show that the amount of radiant energy received in average weather conditions even during the winter months is a significant proportion of what would be obtained with clear skies. The average radiation on north-facing surfaces is always higher than the clear sky value because of the large proportion of diffuse energy from cloudy skies, and in June the energy received on a north-facing surface is 70 % of that incident on a south-facing surface because of the solar altitude angle. East and west-facing surfaces receive on them annually about 60 % of the radiation that would be available for clear skies and south-facing surfaces receive 50 % of their clear sky radiation.

Wilberforce has calculated the net exchange of energy through glazing and has shown that the energy balance has a positive value, i.e. the daily solar gains exceed the average daily conduction losses, over a large part of the heating season, particularly if double glazing is used. Clearly, however, these positive gains extend into the summer months and some means would be required to restrict solar gain through the windows selectively to prevent overheating problems.

The fortuitous solar gains that contribute to the energy balance will not necessarily be useful even in winter and they can be considered useful only if in their absence additional heating energy would be needed to maintain the design indoor temperature. The usefulness of the solar gains depends on a number of factors: time of year, geographical location, orientation and degree of external obstruction, dimensions and thermal properties of the building, glazing area, pattern of occupancy and responsiveness of the heating system to fluctuations in demand. The utilisation of solar gains will vary from one building to another and it is not possible to state a universal utilisation coefficient.

GLARE FROM WINDOWS

Complete uniformity in lighting is dull and unpleasant and can lead to monotony and dissatisfaction. Some brilliance and diversity is usually welcomed but it is necessary to ensure that glare is avoided. Glare is a subjective phenomenon and as such it is an expression of a visual sensation which can not be directly measured. In lighting terms glare is taken to be the expression of the undesirable visual effects which result from the presence within the visual field of areas of excessively high brightness. Two distinct forms of glare can be recognised. They are referred to as 'disability glare' and 'discomfort glare'.

Disability glare is the direct reduction in the ability to discriminate objects or surfaces in the visual field due to the presence of a bright light source such as the sky seen through a window, the light from which becomes scattered in the optic media. This scattered light reduces the contrast sensitivity of the eye and in some circumstances can result in a significant diminution of vision. Disability glare in daylit interiors occurs only in badly lit spaces where a direct view of sky is unavoidable and is more likely to be troublesome on dull days than on bright days.

Discomfort glare is a phenomenon which is more common in buildings and includes the sensations of distraction, annoyance, dazzle and acute discomfort which can result from large areas of bright sky being directly visible from points within a room; it appears to be a compound effect of contrast and saturation.

The degree of glare discomfort has been shown to be a function of:

(i) the luminance of the sky as seen through the window,
(ii) the apparent size of the visible patch of sky, i.e. the solid angle which it subtends at the eye,
(iii) the position of the patch of sky relative to the direction of view, and
(iv) the adaptation conditions in the room, i.e. the average luminance of the room surfaces, not including the window.

These factors are not independent and for example the luminance of the room surfaces is a direct function of the sky luminance and of the window size.

The sensation of glare discomfort is related to these factors in a complicated manner which in daylit interiors appears to suggest that it is influenced primarily by the luminance of the sky and to a very much lesser extent by the area of sky which is visible. Daylight glare is further complicated in that most windows provide a view out on to other buildings or surrounding surfaces the luminance of which is commonly one-tenth that of the sky. Because their location and luminance are intermediate between that of the sky and of the room interior surfaces, the outdoor view can in many circumstances create a buffering effect which reduces the glare discomfort which would otherwise be experienced from direct views of the sky in immediate juxtaposition to the interior surfaces.

A window provides continuity between the indoor and outdoor scenes which register normally in terms of reflectances and colours of readily recognisable objects rather than as luminances or subjective brightnesses. This information content of the view through a window would appear to explain some of the difficulties which have been experienced in experiments designed to assess glare discomfort from windows. Whereas a window of diffusing glass, a roof light or a fogbound scene create visual sensations which correspond to those resulting from artificial sources in 'surface mode', a clear glazed window is seen in 'aperture mode' and there is no precise differentiation between the window as a source of light and its surroundings. Outdoor surfaces and materials are readily identifiable as colours of known reflectances which equate with similar surfaces and materials within the room even though the incident illuminances may differ by one or two orders of magnitude. Only the skilled photometrician or lighting engineer is confounded by this paradox. The man in the street (or more correctly, the man in the room), unencumbered by knowledge of illuminances and apparent brightnesses sees no conflict in the concept of interior view and exterior view as components of a single visual scene.

The window thus extends the visual dimensions of the room and in this respect has significant qualitative advantages over artificial light fittings (luminaires) as sources of light. Windows are seen as part of a room—and a room appears bright and well lit not only because of the amount of light received on the internal surfaces of the room but also because of their relationship with the window and the external surfaces which are visible through it.

REFERENCES

1. HUNT, D. R. G., Personal communication, 1978.
2. 'The IES Code for Interior Lighting,' 1977, Illuminating Engineering Society, London.
3. 'Daytime Lighting in Buildings,' IES Tech. Rpt No. 4, 1972, Illuminating Engineering Society, London.
4. HOPKINSON, R. G., PETHERBRIDGE, P. and LONGMORE, J., *Daylighting*, 1966, Heinemann, London.
5. GRIFFITH, J. W., *Predicting Daylight as Interior Illumination*, 1965, Libby Owens Ford Glass Co., USA.
6. CIE Publication No. 16, 'Daylight: International Recommendations for the Calculation of Natural Daylight', 1970, Commission Internationale de l'Eclairage, Paris.
7. PAIX, D. 'The Design of Buildings for Daylight', Bulletin No. 7, 1962, Commonwealth Experimental Building Station, Australia.
8. 'A Lumen Method for Daylight Design,' Internal Report No. 1/77, Plymouth Polytechnic School of Architecture.
9. 'The Depth of a Daylit Room,' Internal Report No. 2/77, Plymouth Polytechnic School of Architecture.
10. WILBERFORCE, R. R., *Building Services Engineer*, 1976, **43,** 241.

Chapter 8

OPEN SPACES, BLIND ALLEYS

G. P. WOODFORD
B.A., Dip.I.U.
Formerly Senior Planning Officer, Department of the Environment, London, UK

SUMMARY

The Industrial Revolution replaced an immemorial tradition—work in daylight, rest and human contact in the obscurity of home—by a more generalised gregarious darkness both of work-places and homes. A century-long reaction against the dark and overcrowded industrial town has produced a life-style which is different again: the open suburb and undifferentiated artificial lighting of both work-places and homes, a bland environment in which new problems have emerged and the sense of direction and progress is lost. To find a way forward involves re-evaluating sunlight, shadow and obscurity for human life.

INTRODUCTION

This chapter might have been a down-to-earth critical study of the present British rules, regulations and official advice on planning for sunlight and daylight. It is not. Though for several years the author was associated with the formulation of official policy and advice on the subject he is now out of touch, living in retirement in south Brittany. This has advantages. One is unconstrained by conflicting opinions, of specialists or officials, not worried by the fact that the light one actually has to work by comes steeply down over office roofs through dirty windows. One is less concerned with theory but more conscious of light itself—of the fact that the sun keeps the south rooms warm in this mid-season, now the storage heaters are turned off, but leaves the north rooms cold—of the fact that May has been dim and cool this year and has not stimulated the tomato plants to grow or the early potatoes to form. One observes the progress of agriculture, of settlement in the countryside, new houses—daylight coloured—traditional grey and white, scattered across fields and in pine woods. The light of south Brittany when it breaks through clouds on to the sea, sand, granite and trees is usually clear and bright, a reminder that we inhabit a planet of radiant colours suffused from the harsh contrasts of space, a gentle glow of coloured shadows and coloured highlights. In short one is conscious of light not as a factor in the building environment to be provided in specified amounts but as a quality and aspect of life itself. Across the years the insights of boyhood are revived.

There was a time when meadow, grove and stream,
  The earth, and every common sight,
    To me did seem
  Apparell'd in celestial light,
The glory and the freshness of a dream.

It is with such sentiments that I sit down to sort through my own fragmentary experience and offer a personal message more concerned with the character of light for man in his historical predicament than with quantities and techniques. As it is a personal message frequent use of the first person singular seems appropriate.

The fundamental questions relate less to the particular set of standards of natural lighting which affect the shape of settlement today than to the gradual process of adaptation of our way of life and our buildings to the natural environment of light in all its modes. How did we come by the kind of settlement and lighting we have today? Do we get what our eyes and minds and spirits need of natural light and natural obscurity? Are the coming changes in our British habitat and settlement likely to be as responsive to problems of lighting as the changes of the past 100 years? The adoption of particular standards is part of the slow historical process of adaptation and will be considered in this context. In other words I intend to go beyond my own subject, but this, it seems to me, is what we all should do. It involves looking deep into oneself and deep into the past.

LIGHT, PLANETARY ENVIRONMENT AND LIFE

Astronauts must be intensely conscious of the cosmic character of light. Even on earth, protected by magnetosphere and atmosphere, we live in the direct rays of the sun and must sometimes shield our eyes and cover our heads. Icefields, deserts, the open sea, remind us of the power essential to life but from which we must have shelter. The creative conflict between life and radiation is not recent even perhaps on the time scale of the universe; it may have begun in the pre-stellar dust cloud from which it seems the solar system condensed some 4500 million years ago.[1] After some lapse of cosmic time photosynthesis began to fight forward. Thereafter plants commanded atmospheric change, sweetening the air with oxygen, stocking fossil fuels and presumably contributing to a gentler diffusion of light. Life, actually or potentially one of the great forces of the universe, has, so to speak, fought it

FIG. 1. The beds are in cupboards. The traditional 'lit clos', without the step-up coffer in front, is now often used as a side cupboard or a bookcase in bourgeois dining and living rooms.

out on the cosmic time scale and established and organised its place under the fire of cosmic energy; a seedling that moved the rocks. Biology thus stands on a par with physics. The creative conflict continues. We are creatures of light taking our place on a planet precariously prepared for us. Our habitable environment owes the balance of its atmosphere and its resources to plants. We begin to understand that these remain the *sine qua non* of our existence here. Cosmologists are hesitant in their accounts, but such is the version of the genesis parable that I like to accept.

Civilisation began under clear skies and blazing sunlight by the great rivers Tigris and Euphrates, Nile, and Indus. Man became conscious of the sun's periodicity, of the seasons, of annual floods, of cropping timetables, and eventually he learned the sciences of astronomical motions and geometry. Thus a brilliantly lit natural environment and intelligible natural sequences evoked an intellectual response. And also a spiritual response: light and dark, the lightening of darkness, were religious symbols of the millenium of Zoroaster (duality of light and darkness), Buddha (enlightenment), and Christ ('that in Thy light we may see light').

As agriculture spread along rivers into maritime climates, and ships took to continental shelves of the oceans for trade and fishing, livelihood was still gained in the open air; and home was still shelter and shade. Its virtue, like the caves of ancestral hunters, was to be dark, the flame in the hearth illumination and security. The thought brings me back to Brittany where the fishermen's cottages present a blank wall to the west against the weather and the dazzling seascape of evening, and the peasants' cottages open their small windows only to the courtyard on the south, but fireplaces in the gloomy interiors are huge and the beds are in cupboards (Fig. 1). The best places in life look out on the world from the edge of safety—a seat in the sun at the cottage door.

RECENT HISTORY

Progress involved a step into the dark and crowded mills of an industrial society, Britain leading the world. A step from agriculture and fishing under the open sky to *interior* work places and crowded dwellings. A confusion of genres: *then* the darkness of secure homes on the edge of the wide open working world; *now* darkness in both work and home. One is tempted not to dwell on this period, not to look down into the pit of the early 19th century but one has to do so to understand one's own environment. The coal mines: 'women led the horses which drew the trucks, children opened the gates to let them pass. The children began work at the age of seven or eight. Throughout the entire winter they never beheld the daylight. We are told that they could work for 13 hours on end without dying of exhaustion'.[2] 'The miners lived like absolute savages both in the dirty and ruined villages in which they spent the night and in the subterranean galleries where of necessity there was less supervision than in the workshops of a factory.' 'The management of the mine left the payment of the women and children to the miners, together with full liberty to use them at pleasure. . . .' Cotton spinning, in which in 1816 there were twice as many women as men and almost as many children as adults: 'The overseers of the workhouses whose duty it was to lodge the children . . . deposed that they were obliged to keep them under lock and key . . . and obliged also, when work was over for the day, to make the round of the mills to collect those who had fallen asleep in corners. . . .' Cotton weaving to which the factory system arrived much later: 'The weavers worked 'at home' that is to say in an apartment that was not their own—a kennel in which the looms took up all the available room, a dark hovel, almost a cellar: weaving had to be done in a damp place to keep the threads supple.' As steam machinery was introduced the system of two shifts of eleven hours was applied so that half the work was done at night.

In short an incredibly rapid multiplication of underground and indoor work, half lighted, divorced from rhythm of day and night; mill smoke, hearth smoke and smoke of candle and torch obscuring the sky; a window tax that encouraged the building of dark, ill-ventilated rooms.

In our detachment we may suppose that Blake observed the mills in the valley from a hill-top. No, he was inside: it was, I believe, an interior darkness he evoked. Satanic mills: Blake had seen God hunched over chaos with his great dividers separating light from darkness; Satan endeavoured to undo the creation, confuse the issue, nullify distinctions.

Yet most of England remained green and pleasant. Its fertility increased. The rural population grew. The great exodus to the factory towns did not, on the whole, begin to deplete the villages until about 1850. New metalled roads and then the railways opened the way from town to village, from the new environment to the old, from the new generation to the old. As in France today the contact was maintained, urban with rural, town workers with their farmer and fisher parents and relatives. The industrial revolution did not transform the whole environment and its effects were gradually checked by forces both within and outside itself. Peel, a mill owner, forced through the first Factory Act in 1802. Robert Owen, a mill owner, felt himself responsible for working conditions and for housing; he helped to secure government intervention to keep his competitors in line. Outside there were the Methodists and Wilberforce, evangelists and good Samaritan of society. In the muck of the early industrial revolution the seed of a system of social services was planted.

The cultural defence of the countryside and country life was important. Wordsworth, whose themes were nature and light, sensitised intelligent people to their impending loss under the 'rash assault' of the railways and industry. Later the Pre-Raphaelites, friends of the medieval earth, revived nostalgias of old tradition. There were too the painters of nature and light, Constable and Turner. Practical utopians (including O'Connor, a Chartist) founded numerous small, mainly agricultural communities (which inevitably succumbed to the attractions and pressures of the industrial economy). Ruskin's sensitive philosophy was influential. The sense of history and tradition, and love of the countryside and country ways and workmanship were not extinguished. These and the horror of cholera, which appeared like a judgement on the times, (followed by the first Public Health Act in 1846), ensured a reaction.

For the 100 years intervening between then and now the tide of reaction against the social and environmental effects of the industrial revolution has not ceased to flow. The turn round was slow; it was powered by private decisions—the movement of homes and workplaces to the fringes of towns under the influence of high costs and bad conditions in central areas—and facilitated by new modes of local transport. It was driven too by institutional and legal reforms—the setting up of local government as we know it, Public Health Acts and by-laws. Whatever we may think of them today, the early street-width by-laws were a long step towards the salvation of a desperate and chaotic situation. Streets of 36 ft between buildings were a reversion to 18th century practice from before the great darkness (e.g. Lord North Street, Westminster). But soon the enthusiastic legislators prescribed wider and wider separations (in 1912, commonly 50 and 60 ft).

By the end of the 19th century people were beginning to think forwards as well as backwards. What kind of environment was wanted for the future? Ebenezer Howard had a clear and timely answer: an environment composed of man-made and natural, town and country, in equal proportions—towns embedded in country and threaded through with country. It was a religious conception. Garden cities would be like the Kingdom of Heaven: men would love each other and live in the full clear daylight. Unwin, caught up in the practical task of designing garden cities and garden suburbs and eventually recruited into the Ministry of Health, faced the question of the spacing of houses. Should it be 36 ft or 50 or 60?

His definitive answer was 70 ft (Tudor Walters report, 1918) justified directly in terms of sunlight and later, incidentally, in terms of daylight and privacy as well, a multi-purpose spacing standard. Over half a century later it is still one of the most commonly used of housing standards. It allowed for a garage to be fitted into the front garden between house and road; it enabled local authorities to keep builders up to scratch and ensure that the new environment was as open as Unwin intended.

Meanwhile the pure new town ideal did not take on. Everyone felt that 50 % nature was a good proportion and 70 ft an excellent spacing standard, but they wanted to live on the edge of the big town, on a bus route to take them in for work, shopping and school. The inter-war suburban environment, shaped by private enterprise, was a continuation of the 19th century flight from the centre of the town at a different scale and depth, but it was a much lighter and greener environment and often decorated with cherry blossom. The nation took to gardening. Sir Frederick Osborne lightly dismissed this deviant form of Howard's ideal as a 'subsidiary paradox'.[3] In reality Howard and Unwin had released enormous forces which took their own direction imposing a market economy solution. It brought and still brings much private happiness, and in its train serious social and economic planning problems.

Since 1947 settlement formation has been subject to planning control. The majority of new housing has nevertheless continued to be suburban or semi-urban in character even in new towns, so that the flow out from the denser urban areas has continued in spite of considerable dense, inner urban housing construction.

The last quarter of the 19th and the first half of our century thus in great measure achieved the rescue of our living environment. Two successive objectives were clearly perceived and pursued, first to open up and clean up the industrial habitat of the early 19th century, to provide as the legislators of 1875 said 'a sufficiency of space' about buildings; second, to build a new habitat which should be not only adequate but positively desirable, sunlit, with gardens, fit for heroes. The third quarter of our century has however brought disillusion and uncertainty. Suburbs have failed to acquire locally effective social and economic resources, and lack visual character and individuality. Attempts to reconstruct old cities have been questionable. High blocks are now seen as a misguided initiative; town centre developments, in most cases, as destructive of character (particularly of medium-sized country towns); slum clearance and comprehensive redevelopment as socially and economically disruptive. The inner urban areas of great cities, even where redeveloped with high density housing, have lost their multifarious collaborative structure; their population continues to decline. In some areas there is racial separation and racial concentration. Law and order is increasingly threatened. In all this the issue of spaciousness, light and the natural environment of housing is beginning to seem less important in planning than economic and social factors.

Thus we come to the present predicament. I have dwelt at length on the preliminaries, including the extraordinary environmental history of the 19th century, because it seems to me, after working on lighting codes for many years, and now that I can sit and watch the clouds moving rapidly in from the Atlantic and reflect in tranquillity, that we have tended to ignore the delicate rapports which link the visible and spiritual, the past and the future, and thereby to disqualify ourselves from understanding the present.

THE PRESENT PREDICAMENT

The present moment in the history of British settlement is thus a point of great progress achieved; a mainly open and partly natural habitat has taken the place of the intense urban concentration which the industrial revolution was beginning to produce. But it is also a point of growing unsureness, unsureness above all as to the future of the great cities. Howard the initiator of garden housing looked forward with equanimity to their decline and disappearance. He counted on garden cities to reduce their attraction so that ultimately 'the bubble will burst'. But is it conceivable now that we should aim at spacious living for everyone and allow the disintegration of these still populous great city communities to continue and gather momentum?

How have we gone wrong? What has our attitude to natural lighting to do with it? In a sense the wrong turnings have been part of the life history of planners of my generation. One can best refer to one's own experience and that is what I propose to do. It seems to me there are two main strands in our failure. One is a lack of sensitivity and understanding about needs and propensities of men and societies; the other is the loss of clarity as to the kind of settlement that is desirable.

The lack of sensitivity and understanding is manifest in official and expert attitudes to lighting which have remained obstinately quantitative although the quantitative problem is now minor. The industrial revolution which began in Stygian gloom has given us electric light at a touch, picture windows, and by the reaction it provoked a spacious, 50% natural, suburban habitat. But thereby it deprived us of candle and fire flame and variable obscurity within dwellings. Adaptation to bright light outside and obscurity inside is incomparably older than civilisation; it is, I imagine, part of the structure of the mind. Inside cavern or thatched shelter—security, the family, sleep, darkness; outside—daylight, sunlight, the struggle for life and subsistence. The flame of my wood fire in a darkened room is moving by its beauty but surely also by deep unconscious associations. I appreciate its variability, antithesis of the steady brightness of an electric lamp, the parade of shadows it casts compared with which the fluorescent tube casts almost none. Light is the intellectual and spiritual element of mankind; the steady light of a completely overcast sky, or of an electric lamp, a condition of physical existence in which all utilitarian functions are possible. Then the sun comes out. Part of our life is dull without incident, part radiant, full of contrast, shadow and colour. Part of the time we exist, part we live more sensitively and fully. It seems to me that some obscurity, some reduction of the circle of illumination, a move away from the window or the lamp, as well as the final putting out of the light to sleep may be a necessary spiritual comfort. Natural light is continuously variable. The prolonged, almost ritualistic transitions between day and night have a special significance for ordinary people as for poets. The eye is adapted to see in all lights. Is it not almost an insult to subject it to a bland unchanging luminance through long stretches of time? Could it be that the fascination of the television screen owes something to the fact that it replaces the fire in the hearth—a flickering focus?

I do not deny that the quantitative problem still has some importance but I suggest that we have given too little thought to the qualitative aspects of light. Where we do consider it, we tend, it seems to me misguidedly, to increase the blandness of our environment, for example in the concern with avoiding glare. Misguided because we live by glare and take pleasure in it: the blaze of sunlight as we close the door of the house behind us, the blaze of the sea and white sails, of the long snow slope, and, yes, the glare of the small screen in the surrounding darkness.

As a people what kind of environment do we want enough to win and deserve? This is the second strand in our present predicament: we no longer know. We are ambivalent about it: more dense, less light or more open and more light?

The aim was clear in 1880; it was still clear in 1920, and more attractive, but it is not clear now. Both strands, lack of sensitivity and lack of aim can be illustrated by the recent development of the light–spacing–habitat relationship and standards of natural lighting.

Consider the evolution of standards since 1875. During the last part of the 19th century these were simple rules for the width (between buildings) and lengths of road, presupposing a rectangular grid pattern. In place of the chaos of workers' accommodation—shared rooms, tenements, back-to-backs, behind the frontage alley development—they represented order, providing a feasible pattern to accommodate the enormous volume of building required to keep pace with population growth and migration and to reduce the previous multiple occupation and gross overcrowding. In short the standards were the spearhead of a rescue operation at a time of desperate need. They were simple, mandatory, effective.

The rescue operation gave place at the turn of the century to a more forward-looking policy of creating a positively desirable environment. The standard road width between buildings was raised by steps from 36 ft to 70 ft. Before and after World War I settlement spreadeagled into the countryside, mainly on the dense network of main and subsidiary roads on the outskirts of towns. Edwardian terraces gave way to inter-war semi-detached dwellings. The 70 ft rule took firm hold by acquiescence. This was the eminently saleable article, where land had been acquired cheaply and access was ready-made. Consensus rather than mandatory standards created the inter-war suburban environment.

Meanwhile the problem of urban housing and other urban building supervened. How could the third dimension be used? Ideas developed: new architecture, continental theory, specifically the question of light, the 'cité radieuse', a rediscovery of the relation of natural light to urban form (but more theoretical, less pragmatic, than the light and building science of the Georgians) (Figs. 2 and 3). This was the beginning of the analysis of benefits to be sought in the block layout of buildings. The unanalysed enveloping

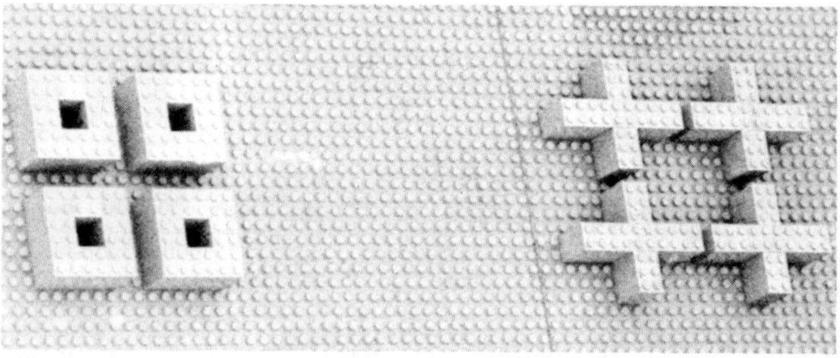

FIG. 2. Rediscovery of the relation of natural light to urban form. Left: simple rules for width and height produced the corridor street, with central light-well inside each block. Right: cruciform block providing the same accommodation but better daylighting.

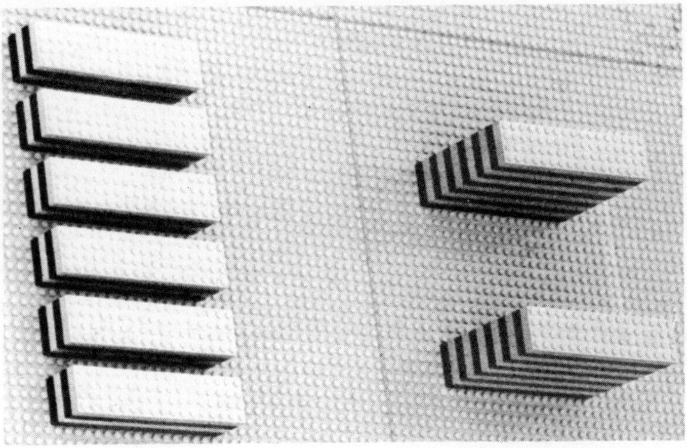

FIG. 3. Planning for daylight. Tall blocks wide apart give more favourable angles of light than low-rise blocks.

concept of 'sufficiency of space' was broken down: space as one element still, but privacy as a separate issue; daylight and sunlight as separate issues; and views out. And there was the architectural idea of enclosure and the related notion of settlements as 'sauvetés' harking back to the ingrained historical human need for security of a community in its settlement as of the family inside the home itself. How to make standards for these complex and subtle interconnected matters? A long journey had been made from 1875 and from simple standards of rescue; but standards still seemed indispensable to further progress.

Then World War II: the 'après guerre' ideal, in contrast with the freely burgeoning ribbons and suburbs of the twenties and thirties, was the deliberately shaped and located settlement of new towns, and elsewhere of positive, formative control. From what could the science and knowledge needed by the first recruits to planning offices and new town development corporations be derived if not from standards? The government set up committees to devise standards and give design advice: for new towns, for central areas, for residential layout and density, for daylight planning and so on. In town and country planning, hitherto a mainly theoretical science, suddenly promoted to be a major practical function of government, standards were an administrative necessity. Many, issued in the form of non-mandatory advice from the centre, were used to justify the mandatory decisions of local government; thus in practice for developers they were mandatory.

To recapitulate: the nature and significance of spacing and lighting standards have changed with the times. Those of the late 19th century were simple, a plain necessity, mandatory and successful in achieving their purpose. Those of the inter-war period were simple too, and were accepted willingly as instrumental in the creation of a spacious and popular suburban environment of houses with gardens. Since 1947 the circumstances have again been transformed, standards are not simple but multiple and complex, and in the matter of lighting, much less necessary than those of the 19th century. Most are not mandatory in theory but may be so in practice, not easily accepted by interested parties, often regarded as frustrating restraints by architects, but felt to be indispensable by those who operate the planning machine. This present uncomfortable state of affairs reflects our disagreements and uncertainty about the kind of settlement we want and how to save the settlement we have.

Natural lighting standards for the formation of settlement were codified for the first time in a MoHLG booklet 'Planning for Daylight and Sunlight' in 1964. This code was intended to be applied at the general layout stage of building design but had the same basis as the BSI daylight code of the same period which gave guidance for the detailed planning of rooms and windows. Underlying these codes were the now questionable daylight assumptions of a completely overcast sky and computation of illuminance on a conventional horizontal surface. The standards were embodied in transparencies by which layout plans could be tested and the position and general shape and height of proposed buildings could be checked. In theory their purpose was to safeguard the two specific benefits of daylight and sunlight. In practice the daylight standards were used to achieve more general, unanalysed benefits of spacing.

In urban areas the code provided valuable reinforcement for policies for the inclusion in layouts of green spaces, car parking areas etc., and for the protection of land destined for future development against the encroachment of buildings, especially non-residential buildings, too near to common boundaries. In London, particularly, the code standards were often the operative method of controlling building density. They had a strong influence on the three dimensional shape of shop and office building, not, I believe, consciously foreseen when the standards were formulated. Podium plus tower block (Fig. 4) or the combination of low with high building were found to be forms which provided satisfactorily both accommodation and standard daylight (a combination of light over the top of the low and round the side of the high).

In rural and low density urban areas the need to protect house property

FIG. 4. The tall block over a low podium was found to meet the daylight code and to satisfy very profitably the commercial demand for shop frontage.

against overshadowing by additional building on subdivided gardens and so on, and from extensions to neighbouring houses often involved references to the 1964 codes which thus strengthened the hand of the preservers of the open environment. However in low density housing layout, the code in conjunction with road width rules and traffic sight lines, tended to produce 'gappiness' in views from houses and footpaths and prevalence of road surface in the foreground of views, particularly at road junctions.

The sunlight standard did not have much effect. Most authorities found it to be over-restrictive (and the transparency difficult to use) and did not apply it.

A revised code was issued in 1971. Looking back on my part in this I now think that the underlying assumption of a completely overcast sky for the computation of daylight should have been changed in this revision, but such a step was then thought to be premature. It would have raised great difficulties for the quantitative assessment of daylight but it would have involved us in some qualitative realism. The character of the sky is to be changeable not uniform. On the other hand it was right, I think, to reinstate the sun. The previously neglected standard for sunlight was relaxed considerably;† the intention was to make it the operative standard for the southerly faces of residential buildings. There was a relaxation too in the housing daylight standard (for northerly faces)‡ since the previous standard had been disproportionately restrictive of low-rise high density building, then beginning to be in favour in urban planning. These relaxations permitting tighter low-rise housing, 'carpeting' with patio houses, the construction of narrower or higher urban residential streets of more like Georgian street section, were in the new spirit of the times. They were an important indication that the long-standing historical trend to ever more open settlement might be reversed.

The 1971 code was deliberately non-mandatory. The need to weigh the benefits of light against other possible but often incompatible benefits was re-emphasised. The standards were shown to be one set of indices, among many others concerned with natural light which it might also be desirable for the designer to consider. The word 'standard' was in fact replaced by 'criterion' which seemed to have a less restrictive meaning. The idea of analysing out the various benefits to be aimed for and the need to assess each separately were advanced. The new code was rather more sensitive to the qualities of light and responsive to architectural opinion and economic pressures. On the other hand it foreshadowed no new firmness or clarity of aim as to the form future settlement should take.

Here I must come back to my own personal feelings about settlement and

† 'Planning for Sunlight and Daylight', HMSO 1971, replaced a standard of potential sunlight of one hour daily during ten months of the year by one of three hours on March 1st. The latter is easier to achieve because the sun's path is higher on March 1st than in late November or late January.

‡ The reference level for the standard was raised by 2 m from the previous ground level reference. This corresponds better to the level of entry of light from the sky into ground floor windows. This relaxation generally increases the permissible height of buildings by about 2 m (Fig. 5).

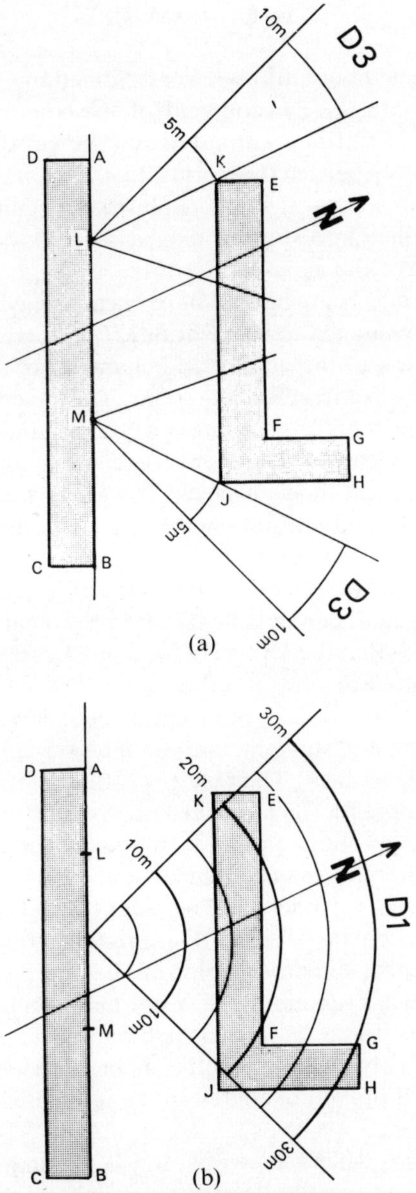

FIG. 5. The massing of adjacent buildings is tested by daylight indicators. (a) Whatever the height of the block EFGHJK, the facade at L and M will satisfy the 'daylight criterion' if the 'wing' of the indicator D3 does not cut the side AB. (b) Indicator D1 shows that the 'permissible height' of the side JK is 13 metres above the reference point.

light and society, and about my own work. Something, it seems to me now, had gone wrong with the development of the country in general in the 'sixties and early 'seventies. Without vision the people perish. The 19th century had vision. Britain led the world and it was felt that what was wrong with settlement and society at home could be put right and British values could even be instilled into subjects overseas. In 1947 the vision of a new world still had force but no clear contours. A huge administration was devised to bring it into being but its shape became less and less clear. Some elements were foreseen: new towns, National Parks, green belts, but there is now no strong vision of the quality and character of settlement we want. Much of what we have is defective, some of it in serious—dangerous—decline. Much of the basis of planning consists of standards, but standards to what end? What ideal? Do they converge towards anything? Who made them? Who can explain their purpose? Towards what fine future can we bend our efforts? The podium and tower to which the daylight standards for non-residential buildings tend: what is fine in that form of settlement, for instance? Do we really want it or was it just the unforeseen product of some inventive planner's mathematical bent? Have the means become ends, the tools our masters? Is that the reverse of the coin? Loss of vision on one side, subservience to standards on the other?

I said my own work. I had been captivated, I confess, by the solid geometry of daylight and sunlight, and though I was a generalist planner, I followed this specialist bent. The result was that I found myself in a side stream where I was practically alone in a huge department of government: no conferring, no meeting of minds of generalist administrator with specialist. The daylighting man could manage. Even most architects would shun the idea of trying to understand or judge the techniques used, or even the solutions to particular problems. This unsatisfactory state may for all I know be the rule; most standards which are based on scientific knowledge are perhaps evolved in a vacuum. To give one man exclusive responsibility is usually expeditious; it may be right if that man is not only a master of his subject but also deeply aware of the needs and tendencies of settlement. Even so he will fail if he cannot confer, in the full sense, about the meaning and aims of his work.

I had no clear idea what was wrong, indeed I clung to what was a not unpleasant, privileged position. Nevertheless when I was working on the 1971 code I suddenly realised that it would have been better to begin with a conspectus of all housing environment standards. Clearly the benefits to be provided: quiet, shelter, pure air, access to facilities, benefits related to roads and so on, were inter-dependent. The provision of one might hinder the provision of another. To make movement of traffic easier might diminish the safety of children and old people; to orientate sitting rooms towards fine views to the north would deprive them of sun. Surely what the designer needed was help in making a *balanced* judgement. Would not a concentration on natural light tend rather to unbalance his judgement, give authorities a weapon of control they could wield abusively. I put it to the Department that the work should be done and that I should like to do it. It was not then possible to spare the inter-disciplinary manpower necessary for the project. Soon afterwards I decided to leave the Department and to seek a grant to organise the research myself. The grant was made and the research carried out.[4]

The result was surprising. I had assumed that standards of settlement together formed a system or network related to the underlying notion of the good environment, and that the object of the research was to show how the holes in the net might be repaired and the net strengthened. The really obvious fact that there was no network but merely a number of separate threads became clear quite soon.† I had unthinkingly felt that standards were derived from a subsisting ideal, a strong, if not clear, vision of the kind of settlement we want. It seems to me now, that such a vision existed at the end of the 19th century and developed during the first half of the 20th century but has now evaporated. Each one of the numerous standards we now have has significance for a particular benefit but none for the coherent whole.

Their incoherence is essentially due to this lack of vision of the whole, to the incoherence of aims. It appears in two ways. First, though numerous, they do not cover the ground: there are no standards for instance for air purity necessary in the residential environment and none for desirable elements in views (green space, sky space etc). Second, each is related only to certain limited aspects of the benefit envisaged. The sunlight standard, for instance, relates only to sunlight in buildings, not to sunlight in gardens (an incompatible benefit in the sense that a ray of the sun can fall on one or the other, not both unless the garden is in or on the building). Privacy standards are concerned only with spacing-derived privacy not at all with screening-derived privacy (which has quite different implications for the form of layout). Daylight standards are concerned with light of a completely overcast sky, not with clear sky of which the luminance distribution depends on orientation, nor with the common condition of variable cloud and multiple chance reflection.

† Not to me but to my collaborator Kirstine Williams.

It gradually became clear to the three of us doing this research that not only were the standards, the threads, not a network but that even if they were multiplied and improved indefinitely they could never, in the absence of a clear articulated ideal of settlement as a whole, form a network in which development could be caught and assessed.

The incoherence of standards is attributable partly to the sudden burst of standard setting activity in the last quarter century. After the end of the war history was moving too quickly: the need to understand settlement seemed less urgent than the need to make prescriptions for it, to feed the hunger of new and unsure teams of planners and designers. Some standards were adapted from traditional models, some were devised in government departments, some researched by specialised bodies (the British Standards Institution, the Building Research Station/Establishment), some formulated by local authorities. Methods of formulation and the research backing had no kind of consistency.

The incoherence of the standards themselves is matched by a discordance of views on how they should be used. Broadly there are three attitudes. Local authority planners regard standards as instruments of power by which decisions can be safely and consistently made. Independently minded designers regard them as a frustrating multiple constraint on the process of creation. The more submissive, law-abiding, home-at-five workers in the field (they are probably the biggest group) are glad to have a set of rules to work by. They then know where they stand and can get on with it.

Looking back on this research job it seems to me that we painstakingly documented a conclusion that should be obvious to any right-minded person. But it had not at first been obvious to us and will surely be treated with scepticism by all those who work for great institutions and who must make or supervise the making of multitudinous decisions, most of all by the Department of the Environment. For what we said in effect is that our present lot of housing environment standards should not be used to make snap decisions for us, but regarded as indices more or less appropriate to the assessment of particular circumstances. Decisions have to be taken in conscience with the available information. However fallible the planners, they must not avoid, by referring to the authority of standards, what is their own responsibility for guiding settlement and society into the future.

As I sit here writing in the garden, conscious of the gradations of green in pine and yew, myrtle, magnolia, sycamore, walnut, acacia, and the grass sunlit and in shadow, another consideration occurs to me, perhaps the most obvious of all, about standards of settlement in general and of natural light in particular. I remember the subject of a sixth form essay at school 'Nature is badly lighted and too green'. Green it certainly is here, and lighted by an intensely brilliant naked lamp radiating half a kilowatt per square metre. But the sun will go in, clouds supervene, the green shift a tone or so. Later the great lamp will dim, redden, sink out of sight, and then darkness. Sometimes, as Bridges said:

> ... grave Night peacefully
> clearing her heav'nly rondure of its turbid veils
> layeth bare the playthings of Creation's babyhood;
> and the immortal fireballs of her uttermost space
> twinkle like friendly rushlights on the countryside.

In other words a slick comment on natural lighting will not do, and for the same reason slick numerical standards are inadequate to resume the values of human settlements or any human affairs. Am I to take it seriously if my son affirms he will choose a wife who (a) smiles twice every five minutes and (b) is not more than 5′ 6″ tall? The simpler the quantities chosen (and for practical purposes they must be simple) the more inadequate they are likely to be. Why is it that when residents are asked how well satisfied they are with the natural lighting of their dwelling on the estate their replies show little or no correlation with the performance of their dwelling in terms of daylight and sunlight standards?[5] Because these standards refer to only two simple conditions: the sun is on (bright sunlight) or off (completely overcast sky). Residents' sensibility is much more sophisticated; they are conscious of, are spiritually sensitive to, bright clouds, sunlit trees, the sun on a white garden wall....

True one pursues reality by successive numerical approximations. The method is a good one, not because one ever catches up but because in the pursuit one perceives gradually more of an infinite qualitative complexity. At various stages of this process indices (rather than standards) may be crystallised out, valuable if they can be intelligently grasped and used discreetly and sceptically by ordinary designers. But more important is sensibility and the truth of imagination: by these we may be guided.

Lighting and associated spacing standards must take their share of the blame for our present predicament. They did not allow the forming of busy commercial–residential streets and tight urban communities essential to urban character and efficient, multiform urban collaborative life. Suburban spacing standards are accused, surely with justification, of producing uniformly characterless and poorly equipped Peacehavens everywhere, pleasant as these may be to live in. The standards of 1875 were clear, purposeful and forceful. Those of today tend to be mysterious to the layman, controversial to the designer, and unrelated to any clear idea of settlement, yet in the hands of local authorities often mandatory routines essential to despatch. Standards contribute to the present predicament not because they are instrinsically bad but because they are badly used. They can no longer point like an arrow the way forward or spearhead coherent policies. Today they can only inform personal judgement and decision and that only if their nature and value is understood.

No one has better indicated the quality required of a designer of settlement than Paul Valéry. His subject is a designer of ships, but the analogy holds. I freely translate from his poetic prose an imaginary discussion between Socrates and Phaedrus.[6]

Socrates Most people, my dear Phaedrus, reason from notions which are not only 'ready made' but which no one has made. No one is responsible for them and so they serve everyone badly.

Phaedrus But he, I tell you, had arrived at altogether personal clarities (clartés).

Socrates That is the only kind that may have universal significance.

Phaedrus He passionately imagined the nature of the winds and waters, the mobilities and resistances of these fluids. He meditated on the generation of tempests and calms, the circulation of tepid currents, and those unmixable rivers that flow mysteriously pure between walls of salt water; he considered the caprices and repentings of breezes, the uncertainties of the deeps and channels, and the treacherous estuaries.

Socrates In God's name how from all that did he make a ship?

Phaedrus He thought that a ship should be in some way created by a knowledge of the sea, and almost fashioned by the wave itself!

DIRECTIONS INTO THE FUTURE

I must try to draw together the threads of what I have written and relate them to the 'uncertainties of the deeps and channels and the treacherous estuaries' of the future. What must we do about natural lighting and the associated form of settlement?

I am sure we cannot create a destiny from pure theory as we fondly supposed in 1947. The standards we have are the débris of the failure of that attempt. Sometimes a man of destiny appears: perhaps Ebenezer Howard was one of them. Yet no new valid aim for settlement can be formulated except in terms of history, the direction being taken by society and changing social values. Planning settlement must be regarded as a contribution to the movement of history. We are what we were and must watch ourselves as we grow from one stage to the next.

There are evident ways in which we are failing to do this. One, I think, is our national over-concern with having a lot of constant light uniformly distributed and with avoiding glare. I would say we need dim as well as bright places in our homes; supper by candlelight, conversation by the fire, sewing, reading, working in the window seat or by a lamp? Surely the long reaction against the darkness of home and work in the early 19th century has gone too far. The characterless suburb, like the bland shadowless fluorescent tube is necessarily unloved. Another example of the neglect of social values: after 1947 house building along existing major and minor roads on the fringes of towns was checked. The reasonably motivated desire to live on a bus route with open fields behind was discounted. A heavy finger on this particular scale prevented the weighing of tendencies that might have held promise if well administered. One cannot make good ships except as a function of the sea. As a maritime people we should realise this.

I can only guess or hope that social pressures will develop in certain ways and I venture to put forward one hopeful guess. A positive result of the inter-war suburb building was to give gardens to millions of people. Plants were man's forerunners and predestined allies on this planet using the energy of sunlight to provide breathable air, fuel, nourishment, materials of construction, shade and shelter; their forms and colours an inspiration of poetry and the visual arts. The symbiosis of man and plant, settlement and cultivation, could be developed and redeveloped in more intimate forms. J. K. Page has pointed out that agricultural operations have increasingly urban characteristics, that for instance the processing of fodder and the disposal of animal waste may be urban scale operations; on the other hand intensive cultivation can be brought into towns and to the home itself. Plants can use human waste products, notably warmth and carbon dioxide. Roofs, conservatories, small greenhouses, provide usable space. Hydroponics attracts interest. Here is a tendency that may be capitalised. Domestic architecture would increasingly take household cultivation into account. The incorporation of the 'greenhouse effect' into homes would relate the growth requirements of plants to domestic solar heating. More attention would be paid to the runoff of rain, to the design of houses for sunshine and shelter, to the drainage of cold air, to local climate and microclimate. Many of these considerations governed country-house building in the 18th century, witness Kenwood in Hampstead and many others, drawing so to speak the forest trees round them for shelter on the north side, the forward southerly sunlit slope draining cold air and rain to a pool. *Le Monde* reports roughly as follows the 'Habitat Solaire' conference in Marseille recently:

'. . . it appears that to construct a solar house is not only to stick a battery of solar panels, of questionable appearance, on to orthodox architecture. It is on the contrary to make a global response to the requirements of thermal comfort; it is to make the whole house a captor of energy by limiting losses, by protecting against wind, by favouring all the bonuses of solar energy through glazing, by the orientation of walls etc; it is in short to adapt the habitat to local climatic factors. This implies, for architecture, not only close collaboration of the heating engineer, but the return to the aesthetic conceptions which take account of a tradition, itself based on experience.'

This should as *Le Monde* says favour local differentiation of architectural styles and tradition. It should in fact surpass the differentiation of traditional styles by creating a closer rapport based on deeper understanding of each of the factors in play including the growth of plants. The effect would not be limited in scale; it would extend to the designing of settlements. Microclimate is no less important in towns than in the countryside; wind problems may be worse, and trees as windbreaks no less necessary. The landscape of towns would become greener, vegetation more luxuriant and ubiquitous, especially in London, climatically the most favoured of great cities in Britain.

Our historical predicament is to have failed to evolve a pattern of settlement and habitat appropriate to an industrial civilisation. In the first part of the 19th century the failure was complete. Industrial urban conditions were desperate. In the last 100 years a sustained attempt to recreate a balance between urban and rural has taken us a long way back from the 19th century industrial town towards a natural environment. In the now depopulated and endangered inner cities, however, the imbalance remains. I can see no better hope for the longterm future than a continuation of the movement which had its origin in the idea of Howard and Unwin at the turn of the century—the garden settlement. The idea has not lost its power and may now find new forms of expression, a re-integration of nature with settlement on the scale of the home and the farm as of the town and the countryside, a rediscovery of the natural fertility of cities and of the fact that plants will grow in close conjunction with people. One may hope that the application of the sciences of light, solar energy, climate, and plant growth to settlements and homes will in time tend to assimilate towns with suburbs and farms and the natural environment, not exactly in the way Howard envisaged, nor in the suburban spirit of the inter-war years, but more completely and intimately.

Solar radiation is the element in which the seed of life on our planet germinated. I believe, as Wordsworth did, that our feeling for natural light is a feeling for life itself. I have tried to show that one aspect of history is man's developing attitude to light, and I have picked out from recent social history a particular strand, or rather a strongly flowing stream, the movement towards garden settlements, which gives hope for the future. The movement depends for its impetus on its appeal to people and will not be much encouraged by standards or specialist opinion. To make a good judgement about lighting one cannot abstract light from things seen in the light, nor from the instinctive needs of the seer, nor from the complexity of social conditions. On the other hand if we all speak from heart and instinct about the kind of place we want to live in, we may eventually find we agree with each other and begin to deserve and attain it.

REFERENCES

1. WICKRAMASINGHE, C., *New Scientist*, 1977, **74**, 119.
2. HALÉVY, E., *A History of the English People in* 1918, 1937, Penguin, Harmondsworth. This and subsequent quotations are based on contemporary records and reports of parliamentary enquiries.
3. OSBORNE, F. J., introduction to Howard, E., *Garden Cities of Tomorrow*, 1945, Faber, London.
4. 'The Value of Standards for the External Residential Environment', 1976, Department of the Environment, London.
5. Detailed survey of sunlight and daylight conditions in 47 dwellings on six council estates by Sociological Research Section, Development Group, MoHLG, and measurements by Pilkington's Environmental Advisory Service, 1969, referred to in the forthcoming BSI Code for Sunlight.
6. VALÉRY, P., *Eupalinos ou l'Architecte*, 1934, Gallimard, Paris.

INDEX